SpringerBriefs in Complexity

SpringerBriefs in Complexity are a series of slim high-quality publications encompassing the entire spectrum of complex systems science and technology. Featuring compact volumes of 50 to 125 pages (approximately 20,000-45,000 words), Briefs are shorter than a conventional book but longer than a journal article. Thus Briefs serve as timely, concise tools for students, researchers, and professionals.

Typical texts for publication might include:

- A snapshot review of the current state of a hot or emerging field
- A concise introduction to core concepts that students must understand in order to make independent contributions
- An extended research report giving more details and discussion than is possible in a conventional journal article,
- A manual describing underlying principles and best practices for an experimental or computational technique
- An essay exploring new ideas broader topics such as science and society

Briefs allow authors to present their ideas and readers to absorb them with minimal time investment. Briefs are published as part of Springer's eBook collection, with millions of users worldwide. In addition, Briefs are available, just like books, for individual print and electronic purchase. Briefs are characterized by fast, global electronic dissemination, straightforward publishing agreements, easy-to-use manuscript preparation and formatting guidelines, and expedited production schedules. We aim for publication 8-12 weeks after acceptance.

SpringerBriefs in Complexity are an integral part of the Springer Complexity publishing program. Proposals should be sent to the responsible Springer editors or to a member of the Springer Complexity editorial and program advisory board (springer.com/complexity).

More information about this series at http://www.springer.com/series/8907

Andrzej Nowak • Robin Vallacher
Agnieszka Rychwalska
Magdalena Roszczyńska-Kurasińska
Karolina Ziembowicz • Mikołaj Biesaga
Marta Kacprzyk-Murawska

Target in Control

Social Influence as Distributed Information Processing

 Springer

Andrzej Nowak
Department of Psychology
Institute for Social Studies
University of Warsaw
Warsaw, Poland

Agnieszka Rychwalska
The Robert Zajonc Institute for
Social Studies
University of Warsaw
Warsaw, Poland

Karolina Ziembowicz
The Maria Grzegorzewska University
Warsaw, Poland

Marta Kacprzyk-Murawska
The Robert Zajonc Institute for
Social Studies
University of Warsaw
Warsaw, Poland

Robin Vallacher
Department of Psychology
Florida Atlantic University
Boca Raton, FL, USA

Magdalena Roszczyńska-Kurasińska
The Robert Zajonc Institute for
Social Studies
University of Warsaw
Warsaw, Poland

Mikołaj Biesaga
The Robert Zajonc Institute for
Social Studies
University of Warsaw
Warsaw, Poland

ISSN 2191-5326   ISSN 2191-5334 (electronic)
SpringerBriefs in Complexity
ISBN 978-3-030-30621-2   ISBN 978-3-030-30622-9 (eBook)
https://doi.org/10.1007/978-3-030-30622-9

This Springer imprint is published by the registered company Springer Nature Switzerland AG.
The registered company address is: Gewerbestrasse 11, 6330 Cham, Switzerland

Foreword

Traditionally, the theories and methods of the social sciences and humanities—the so-called "soft" sciences—were believed to be irreconcilable with the perspective and methods of physics, chemistry, and the like, the so-called "hard" sciences (Snow 1959). The bridging of these distinct domains of inquiry within a single conceptual framework in recent years counts as one of the greatest achievements of the complexity perspective. Theoretical concepts, analytical tools, and research methods that originated in physics, mathematics, and computer science, including such notions as emergence (e.g., Holland 2000), self-organization (e.g., Ulrich and Probst 2012), complex networks (e.g., Watts and Strogatz 1998; Albert and Barabási 2002; Strogatz 2001), complex systems (e.g., Sawyer and Sawyer 2005), dynamical systems (e.g., Nowak and Vallacher 1998; Vallacher and Nowak; Devaney 2008), and agent-based modeling (e.g., Davidsson 2002; Gilbert 2004), have revolutionized psychology, the social science, and the humanities.

The approach of complex systems, however, has also generated the potential for influence in the other direction, such that insights generated in the social sciences may inform theory and research in the physical sciences. Armed with the tools developed in physics, mathematics, and computer science, social scientists can investigate questions and explore avenues that are of interest to theorists and researchers in the hard sciences. Indeed, human groups and societies are increasingly recognized as arguably the most complex of all systems and thus represent a new challenge for those who heretofore had focused on physical phenomena (Vallacher et al. 2018). Human social systems are characterized by elaborate structure; they are capable of coordinated action and have their own culture, goals, norms, and values. These features of groups and societies reflect the interactions among individuals who collectively gather and process information, manage knowledge, construct a shared reality, converge on common attitudes and judgments, and collectively make decisions. In fact, the most astonishing achievements of humanity, culture, and science have been created in this process.

Building on this perspective, we argue in this book that human groups and societies are complex, distributed information processing systems. The quality of information processing in these systems directly translates to the well-being of social groups and societies. The importance of the social aspects of information processing is also clearly visible on the level of an individual. When learning about new facts, interpreting and integrating knowledge, arriving at a judgment, or making a decision, individuals seek others for information, interpretation, and advice. Even if individuals reach an independent decision, they confirm their decision by consulting others. They also learn from others how to select, integrate, and evaluate information. The crucial ability to involve others in one's own information processing is determined by the individual's social capital (Bourdieu 1997) and group success (e.g., Coleman 1988; Putnam 2001; Fukuyama 2002).

The perspective of socially distributed information processing has not been lost on modern computer science, as reflected, for example, in the research on automatic recommendation systems. Our goal is thus to use the knowledge of social psychology to understand how the processes of distributed social information processing operate at the individual and the group level. We argue that the rules of social influence, which have received considerable attention in social psychology, describe how distributed social information processing works.

Warsaw, Poland	Andrzej Nowak
Boca Raton, FL, USA	Robin Vallacher
Warsaw, Poland	Agnieszka Rychwalska
Warsaw, Poland	Magdalena Roszczyńska-Kurasińska
Warsaw, Poland	Karolina Ziembowicz
Warsaw, Poland	Mikołaj Biesaga
Warsaw, Poland	Marta Kacprzyk-Murawska

Acknowledgment

This work was supported by funds from Polish National Science Centre (project no. DEC-2011/02/A/HS6/00231).

We want to thank Jeremy Pitt for his helpful comments on this manuscript.

Contents

Chapter 1
Social Influence as Socially Distributed Information Processing

The capacity for thinking is what defines our species—*homo sapiens*. Under this assumption, people are equipped by evolution to think for themselves and thus solve problems and satisfy their needs. However, an equally basic feature of our species is our sociality. Indeed, humans function best when relying on one another to solve problems and satisfy their needs. In the absence of sufficient knowledge or confidence in one's judgment, people will seek out others for their opinion. Sometimes even when the transaction costs involved in the consultation are higher than the apparent gain in the outcome of the decision. Even concerning the most mundane issues (e.g., which paper clips to purchase), an individual may seek out information and advice from others in order to make an informed decision. This reliance on social sources of information is especially pervasive when the issues are complex and ambiguous, and the information is doubtful. Choosing a career path, for example, is an impactful decision that requires the integration of many different considerations that might overwhelm the time and cognitive resources of an individual. Beyond the inefficiency in attempting to process all the relevant information by oneself, the quality of the resultant judgment or decision may also suffer. So when an individual must make a decision in an information-rich and complex environment or judge the relative value of different courses of action, he or she looks to others to provide relevant information and to assist in the integration of such information. In effect, then, the individual delegates at least part of the judgment and decision-making process to other people. Essentially, this is tantamount to people's willingness to be influenced by others, even if this means relinquishing some degree of personal control and autonomy.

However, reliance on others for the gathering and integration of information is associated with potential risks. Are the opinions and recommendations of others worthy of embracing? Can the source of information be trusted to provide accurate and helpful information rather than attempting to advance only his or her personal and hidden agenda? Navigating the intricacies of reliance on other people while avoiding the risks of doing so requires a high degree of intelligence on the part of

A. Nowak et al., *Target in Control*, SpringerBriefs in Complexity,
https://doi.org/10.1007/978-3-030-30622-9_1

1

individuals. The readiness to be influenced, in other words, is not automatic but instead often involves difficult decisions and judgments. The theory we propose focuses directly on this dilemma of social life. It outlines the fundamental processes by which people decide when to delegate information processing to others versus undertaking information processing on their own. If information processing is delegated to others, what determines which part of the information processing is delegated? When do people delegate the gathering of basic facts and information, but reserve for themselves how this information is integrated to reach a judgment or a decision? When do people instead delegate to others the entire process of decision-making, including conclusions and action recommendations?

1.1   Overview

When information processing is delegated to others, each individual in a social system is both the *source* and *target* of influence to varying degrees. In the context of groups, people interact to exchange information and opinions, express their views, and obtain advice from others. From this perspective, social groups can be considered *distributed information processing systems*. The rules by which individuals optimize their cognitive functioning through the delegation of information processing have consequences at the group level, such that it enhances the functioning of the group as well. Indeed, the significant innovations and achievements in art, culture, science, and technology have all resulted from collective efforts rather than merely the insights and efforts of individuals (e.g., Johnson 2009). Although individual attitudes and accomplishments are not to be dismissed, their primary importance is the construction of a *shared reality* that goes beyond the sum of each individual's contribution. This shared reality then becomes the platform for subsequent group judgments, decisions, and actions. The quality of knowledge that underlies the construction of shared reality is the key to the success of the group's actions. Systematic distortions of knowledge, such as fake news or efforts to undermine the group's legitimate knowledge, can significantly impair the effectiveness of group decisions and actions.

In this view, social groups can be conceptualized and investigated as *complex systems* (e.g., Arrow et al. 2000; Nowak et al. 1990). The members of a group are the elements and rules of social psychology, and other social sciences, describe how the elements interact with one another to produce group-level properties. Many of the rules describing how elements in a social system interact have been studied in social psychology as principles and mechanisms of social influence (e.g., Allport 1954; Cialdini 1993). Although this approach has produced many valuable insights and empirical generalizations, social influence for the most part has been viewed from the perspective of the source of influence, where the main questions center on how the source can change and manipulate the attitudes, beliefs, and actions of a target. This essentially equates social influence with power and control, and gives the information advantage to the source rather than to his or her targets.

In this book, we establish a level playing field between the source and target of influence, such that both play important roles in the social influence process, and thus in the creation of a shared reality for the social system in which they are embedded. Indeed, what looks to be the target of influence often initiates the influence process, selects the source of influence, and chooses the type of the influence. Extending the understanding of social influence in this fashion enables one to use the rules of social influence to understand how socially distributed information processing occurs at both the level of the individual and the social system (i.e., groups and societies). Especially important in this regard are *trust*, *coherence*, and *importance* as deciding factors that control the delegation of information processing. These factors have played prominent roles in traditional theory and research on social influence, but they were typically viewed from the perspective of the influence agent and thus are not directly applicable to the description of socially distributed information processing systems.

In leveling the playing field between the source and target new questions emerge. Who is chosen as the source of influence by the target? What type of influence is sought (e.g., facts vs. judgments and decisions)? When does the target seek influence rather than gather and process information on his or her own? More broadly, viewing social influence as socially distributed information processing enables one to consider the optimality of social influence processes. When does the delegation of information processing result in effective decisions and accurate judgments versus bad decisions and inaccurate judgments? Apart from the resultant quality of socially distributed processing, this new perspective enables one to consider its efficiency. When is the delegation of information processing to others less costly, both in terms of effort and risk, than is the processing of information by oneself? On the group level, what rules of allocation of information processing to group members lead to higher efficiency of group-level processing in terms of optimal utilization of group resources (e.g., individual expertise, reliability, and availability of time, minimizing the possibility of information processing failure)?

1.2   The Canonical View of Social Influence

Social influence is the most ubiquitous process concerning the interactions of individuals in social groups and society generally. Asch (1955) defined social influence almost as broadly as the field of social psychology, as any change in emotions, thoughts, and behavior resulting from the real or imagined presence of others. With this definition in mind, social influence underlies virtually all phenomena in social psychology, from attitude change and persuasion to close relationships and the formation of public opinion in society.

A variety of social influence mechanisms underlie central topics in social psychology: conformity (Asch 1956), attitude change (e.g., McGuire 1985; Sherif et al. 1965), persuasion (Petty and Cacioppo 1986), obedience to authority (Milgram 1974), social power (French et al. 1959) and many more (cf. Nowak et al. 2003;

Wojciszke 2000). Social influence also forms the basis of coordinated social action, group formation, and social bonding in general (Grzelak and Nowak 2000).

The vast majority of research on social influence has been conducted from the perspective of the source of influence, with a focus on how the influence source attempts to control the target's thoughts, feelings, and behavior (e.g., Cialdini 1993). Under this assumption, social influence is tantamount to power and thus typically is operationalized and investigated in terms of manipulation strategies and techniques. Influence is implicitly assumed to act against the will or even the interests of the target of influence. For example, an influence agent may attempt to persuade a consumer to purchase a car that he or she would not otherwise buy. Social influence can also work against the intentions of the target. For example, an influence agent may attempt to influence a target to quit smoking. On the societal level, the issue is how can an influence agent induce a desired social change, such as public opinion regarding environmental awareness, government regulations, taxation, or the adoption of a new product.

When viewed from the influencing agent's perspective, the target of influence either bends to the influence attempt or manages to resist the influence. Accordingly, research to date from the target's perspective has emphasized psychological reactance (Brehm 1966)—the tendency to resist influence and maintain a sense of personal freedom and autonomy. Although people certainly do resist influence, they often find influence to be personally desirable. A manipulation attempt will be much more effective when the influence target desires to be influenced.

In more general terms, the canonical view of social influence can be enhanced by insights forthcoming from recent adaptations of principles of complex systems to social psychological phenomena (e.g., Arrow et al. 2000; Guastello et al. 2008; Nowak et al. 2013; Nowak and Vallacher 1998; Vallacher and Nowak 1994, 2007; Vallacher et al. 2002). First of all, this perspective emphasizes the frequent bidirectional causal influence between variables, in contrast to the traditional approach in social psychology which distinguishes between causes (independent variables) and their effects (dependent variables). In reality, as stressed by the complex systems perspective, the relations between variables operate in terms of positive and negative feedback loops, such that each variable both influences and is influenced by the other variables. Second, the feedback loops are not self-contained and isolated from one another, but rather they form a dynamical system in which the individual feedback loops combine to define the properties and functioning of the system as a whole. For example, the rules of interaction between elements define dynamical patterns of change at the level of the system. Systems can thus be characterized by patterns of temporal changes, stability, attractors, and other characteristics of complex systems. Third, the laws governing the group-level may be different from laws governing the dynamics of elements and their interactions, and cannot be easily reduced to dynamics at that level. This phenomenon is called *emergence*. For example, very simple rules of interactions among elements may result in very complex properties for the system as a whole (e.g., Johnson 2001; Nowak 2004).

1.3   Social Influence as the Delegation of Information Processing

When we examine real-life examples of social influence, often a very different picture of social influence emerges. Individuals often desire and seek to be influenced, when they lack the knowledge to make a decision or action, or to save their own efforts in finding and processing information. For example, people often make decisions based on the observations or explicit recommendations of other people. When trying to figure out how to purchase a train ticket from a machine in a foreign country, we observe how other people solve this problem. We tend to buy products that our friends have chosen for themselves. At an exhibition, we follow the crowd to find the most interesting artifacts. We let doctors decide how to cure our diseases, ask dieticians what we should eat, and ask priests how we should resolve moral dilemmas. Individuals ask others for directions and advice when deciding which products to buy.

1.4   The Ubiquity of the Delegation of Information Processing

Being open to influence is ubiquitous among all social animals. Among herd animals, for example, one member of the herd functions as the leader, deciding where to graze and how to protect against predators. Contrary to conventional beliefs and intuition, however, the leader is not necessarily the strongest member of the herd but rather the "wisest" member who is best able to model or dictate what the herd should do. In wild horses, for example, it is the wisest mare, rather than the strongest stallion, that emerges as the leader and takes on the responsibility of being vigilant regarding danger and being sensitive to food sources. In one sense, the leader of the herd is in a position of power, but in another sense, he or she experiences a burden that other members of the herd are spared. In fact, the burden of leadership can promote stress and anxiety (e.g., Brady 1958), whereas the seemingly powerless "followers" can avoid stress and go about their lives with less anxiety.

Social influence in the form of the delegation of information processing is ubiquitous in social life. Whether the issue is trivial (e.g., where to sit at a ceremony, how to tip in a restaurant) or consequential (e.g., purchasing a house, choosing a surgeon for an operation), individuals routinely look to others for cues concerning how to behave, and for information and advice regarding judgments and decisions. They delegate decision making to others perceived to be competent or they use the information and advice they get from others when making their judgments and decisions. Thus, in social groups, everyone is both a source and a target of influence.

In social groups, everyone is both a source and a target of influence. In this view, the group is a complex parallel processing system based on optimizing principles. By spreading information processing to all the group members, the group may process more information in an efficient manner that benefits everyone. By delegating the processing to those who are most qualified, the quality of the information processing by the group is optimized. The ubiquity of social influence thus enhances an individual's ability to function, and this enhances the functioning of the group.

1.5   The Benefits of Delegation for the Influence Agent and the Influence Target

In searching for influence, individuals observe others and seek information and opinions from them. Social influence processes enable the targets of influence to utilize the knowledge and processing capacities of sources of influence to reduce their effort and optimize their functioning. This process is reminiscent of *transactive memory* (Wegner 1987), where the individual relies on the memory of others to store information so that he or she need only remember who has the relevant information. Similarly, social influence from the target's perspective represents transactional information processing, where individuals delegate different aspects of information search and the formation of opinions, judgments, and decisions to others.

Because the choice to be influenced is tantamount to the delegation of information processing, it is beneficial to the influence target. The target, first of all, is able to use his or her cognitive resources for other tasks. Second, if the target chooses an appropriate source of influence (e.g., one who is highly competent), this is likely to result in a higher-quality decision, judgment, or opinion that is a decision made on one's own. Letting an expert choose a car or a camera, for example, is likely to result in purchasing a better product than a product chosen by oneself. In this vein, Ormerod (2015) argues that in the modern world there are so many competing and complex products that it is virtually impossible for individuals to engage in rational decision making by considering all the relevant features of products. Hence, it is efficient and beneficial for individuals to adopt the choices and decisions of other people. Delegating influence thus optimizes the functioning of individuals. The capacity to delegate the gathering or processing of information to appropriate others is a manifestation of social capital, while the necessity to collect and process the information oneself often is a sign of social isolation rather than power and independence.

Consider the example of consumer behavior. In today's world, individuals are overwhelmed with both the number and the complexity of products available for choice (Bentley et al. 2011). In Manhattan, for example, there are 50,000 restaurants, which makes it virtually impossible to make a fully informed and rational choice of where to go for lunch or dinner. As another example, choosing between high-tech products such as smartphones or cameras requires the decision-maker to have technical knowledge much above the level of the average consumer. In both cases, the decisions are made by relying on the information processing already done by others, or by delegating information processing or even decision making to others.

From this perspective, the capacity to delegate information processing and decision making to others can be viewed as a sign of the target's power. Indeed, one could argue that the target of influence is in more powerful position than is the source. After all, when the target delegates to the influence agent, he or she is effectively relieved of the task of information gathering and integration, leaving these time-consuming and effortful actions to the influence agent. If the influence agent is an expert, the effortful nature of his or her responsibility may be fairly minimal. However, if the influence agent lacks relevant knowledge and expertise, he or she may feel the burden of the delegation process because this process is likely to require seeking out information that is not currently available to him or her. Beyond that, providing good advice usually requires an understanding of the needs and intentions of the target, and thus sensitivity to the target's current state, situational constraints, and standards for acceptable information. So rather than feeling powerful in relation to the target, the influence agent must accept the interpersonal burden of being polite and acting in a responsible manner—actions that are often more closely aligned with subordination than with power.

Of course, there may be benefits to both the target and source of influence in the delegation process. Although the influence agent is charged with the effortful and time-consuming task of gathering and integrating information, he or she has decision control over the target. In addition to feeling competent and dominant, the influence agent benefits by virtue of satisfying the basic human need to be needed in a social relationship. Theory and research have documented the ubiquity and power of the need to belong (Baumeister and Leary 1995) and the importance of the synchronization of behavior and internal states in social relations (Nowak et al. 2017; Nowak et al. 2020). In this sense, the delegation of information processing and decision making captures the essence of social exchange and reciprocity in interpersonal relations (e.g., Gouldner 1960; Thibaut and Kelley 1959).

1.6 New Questions Concerning Influence

By emphasizing the targets' perspective regarding social influence, different questions become central for an understanding of influence that differ from the questions that arise from considering the influence agent's perspective. Rather than asking how to maximize influence and control of the thought, feelings, and behavior of the target, one asks how an individual optimizes his or her functioning using influence from others, in what situations the influence of the source is beneficial for the target, and how to overcome potential traps related to the delegated processing. Of central importance are the following questions: (1) whom does the target seek out for influence? (2) what type of influence does the target seek? (3) how is information from different sources combined?

1.7 Regulatory Theory of Social Influence

The above questions are addressed by the *Regulatory Theory of Social Influence* (RTSI). This term is employed to emphasize the optimization made possible by the rules of social influence. In particular, both the quality and efficiency of information gathering, judgment, and decision making in social systems are enhanced by delegating parts of the process to others.

Our theory integrates significant contributions of important theories in social psychology that describe the social influence process. These theories include Action Identification (Vallacher and Wegner 1985), Elaboration-likelihood (Petty and Cacioppo 1986), Unimodel (Kruglanski 1996; Kruglanski and Thompson 1999), Transactive Memory (Wegner 1987), as well as the Dynamical Theory of Social Impact (Nowak et al. 1990). Our approach expands these theories by precisely specifying the mechanisms that must underlie them, as well as introducing new factors that modify the effects proposed by these theories. The proposed theory bridges individual and group levels of social influence by showing how micro-rules produce a distributed social processing system that enables groups to optimize their function.

1.7.1 Primary Factors in RTSI

RTSI specifies four primary factors that individually and collectively determine the choice of an influence source: the *importance* of the issue, *trust* in the source of influence, the *coherence* of the source, and the individual's *expertise*. These factors and their interactions dictate both the degree to which information processing is delegated to others and the stage of information processing that is delegated. Below we consider these factors and their influence on delegation and the level of information delegation. After doing so, we will consider how these factors interact with one another in specific ways to determine the nature and extent of delegation in information processing.

1.7.2 Importance

People think differently about an issue depending on the issue's importance. Importance, of course, is a subjective matter—what is essential to one person could be trivial to another—making the objective nature of importance difficult, if not impossible, to determine. The issue, then, is what makes an issue subjectively important. An essential feature of subjective importance is the issue's *personal relevance*—whether the issue has consequences for the individual (Apsler and Sears 1968). This is obvious with respect to rewards and costs—the larger the reward or, the greater the cost associated with an issue, the higher the importance. Such consequences are not confined to those that are physical (e.g., safety, health, rewards, and costs), but also include consequences that reflect on one's personal characteristics (e.g., personality traits, values) and overall self-evaluation concerning basic dimensions of evaluation such as morality and competence.

A closely related aspect of importance is the degree to which the issue reflects on the individual's *social identity*. An issue that may not have direct relevance to the individual's personal concerns may nonetheless be viewed as highly important if it has perceived consequences for the group or category with which the individual identifies him or herself. Having a particular religious or ethnic identity, for example, sensitizes the individual to issues that have consequences for that particular identity.

A third factor dictating an issue's subjective importance is its *physical or temporal distance* (e.g., Liberman and Trope 2008; Liberman et al. 2007; Stephen et al. 2010). If the issue concerns an event that is physically proximate to the individual (e.g., a crime incident) or that is likely to occur in the immediate future (e.g., a severe storm), the individual is likely to attach correspondingly high importance to the issue.

Issue importance plays two fundamental roles in RTSI. First, it determines, in large part, whether the information processing is delegated to others or is done by the individual him or herself. The less important the issue, the more likely the individual is to adopt the opinions, decisions, and recommendations of others. In this case, the individual automatically follows the lead of others and does so without making a conscious decision. Deciding how to move around an obstacle, for example, is simply a matter of doing what others do when confronted with the obstacle. Similarly, a trivial purchase (e.g., paper clips) is unlikely to engage the individual's conscious decision making. However, when the issue is important, the individual does not blindly model the behavior of others. In deciding to purchase an automobile, for example, the individual is unlikely to be swayed solely by the purchasing decisions of others, but instead will gather relevant information on his or her own.

Importance not only dictates whether we delegate information processing to others but also what part of the information process is delegated. For relatively unimportant issues, the individual is likely to seek *high-level information*—that is, information that is has been processed and is presented in terms of conclusion, recommendations, judgments, decisions, and so forth. This has the benefit of conserving the individual's cognitive resources, which are then available for other tasks of greater importance. At the same time, though, such delegation effectively deprives the individual of control over the information integration process. For issues with greater importance, on the other hand, the individual is likely to delegate only the gathering of *low-level information*—the basic facts and events relevant to a judgment--reserving for him or herself the processing and integration of the information, which is manifest as a decision, judgment, or action recommendation. The difference between objective facts (i.e., low-level information) and subjective opinions (i.e., high-level information) is critical for the obvious reason that objective facts can be characterized as true or false and thus are open to verification, whereas subjective opinions cannot be as easily assessed with respect to their truth versus false value. Surprisingly, though, the objective-subjective distinction has only rarely been considered in the context of social influence and opinion dynamics (Baron et al. 1996; Giardini et al. 2011).

These points regarding issue importance are consistent with basic tenets of the *Elaboration Likelihood Model* (Petty and Cacioppo 1986) and the *Unimodel* of influence (Kruglanski 1996). Central to both models is the elaboration continuum, the ends of which correspond to two routes to persuasion. *Central route processes* are those that require a great deal of thought, and therefore are likely to predominate under conditions that promote high elaboration. Such elaboration involves careful information processing of relevant facts and considerations rather than relying on the source of information to interpret and evaluate these facts and considerations. *Peripheral route processes*, on the other hand, often rely on contextual characteristics of the message, such as the perceived credibility of the source (Heesacker et al. 1983), the attractiveness of the source (Maddux and Rogers 1980), or the catchy slogan that contains the message. In this case, the individual is likely to rely on the source to integrate the information and generate the corresponding conclusions. The attention to peripheral cues decides about the individual's acceptance of these conclusions.

The Unimodel shares this basic scenario but assumes that the distinction between central and peripheral routes is not categorical but rather is proportional to the amount of cognitive effort devoted to the rational processing of information. The specific mixture of central and peripheral processing reflects how deeply versus shallowly the arguments and information regarding the issue are processed or the proportion of arguments that are rationally processed. It should be noted, though, that even for important issues that might otherwise be processed rationally, a limitation in cognitive resources may interfere with central processing and promote primarily peripheral processing. By the same token, issues that are perceived as difficult and thus require central processing are more likely to be processed through the peripheral route. Of course, there are individual differences in the tendency to process issues through the central versus peripheral route. People high in *need for cognition* (Cacioppo and Petty 1982) and low in *need for cognitive closure* (Kruglanski and Webster 1996), for example, are more inclined to process even unimportant and simple issues in a rational manner than are those with a low need for cognition and a high need for cognitive closure.

The distinction between central and peripheral processing is associated with how strongly an attitude is held and has implications for attitude stability (Krosnick 1988; Schuman and Presser 1981) and attitude changes (Fine 1957; Gorn 1975). In particular, attitudes acquired through the central route have greater persistence over time and are more resistant to attempts to change them through counter-arguments (Petty and Cacioppo 1986).

If important attitudes formed through the central route do change, the pattern of change is qualitatively different from the pattern of change of attitudes formed through the peripheral route (Latané and Nowak 1994; Nowak and Vallacher 1998). For relatively unimportant issues, change may be incremental and in proportion to the strength of the influence. Attitude change concerning important issues, in contrast, follow a catastrophic pattern, with long periods of resistance followed by abrupt and wholesale change. So although attitudes on unimportant issues change in a linear fashion, attitudes on important issues change in a nonlinear fashion. Catastrophe theory (e.g., Thom 1975) describes the transition between the linear and nonlinear patterns with changes in the issue's importance.

1.7.3   Trust

A second primary factor is *trust*. Whereas importance mostly dictates whether people seek influence from others rather than investing their own resources, trust dictates which source of influence is chosen (e.g., Riegelsberger et al. 2005). People prefer to obtain from those they trust, and they are likely to follow such advice. There are several aspects to trust. First, the source of influence must be judged to have *good intentions*, so that the information or decisions he or she provides is unlikely to be offered in an attempt to harm the target. Second, the source of influence must be judged as *willing to help* the target's information gathering and decision making. Third, the source of influence must be judged as *competent to help* the target by providing relevant and accurate information and providing sound judgments and decisions.

When trust is low, individuals are unlikely to delegate the whole decision-making process; instead, they are likely to retain some control over the process. In doing so, they are likely to delegate those parts of the information processing that are more objective, less subject to bias, and easier to check. This represents what is often called the "trust but verify" strategy. So the individual is likely to delegate only the gathering of basic factual information to the influence agent, not the decisions and judgments. When trust is high (e.g., the influence is judged to be both well-intentioned and highly competent), however, the individual is likely all information processing to the agent, including the decisions and judgments regarding the issue.

In principle, the individual should seek influence from the most trusted source of influence. In most cases, however, this may not be a viable option because the available sources of information may not be as trustworthy as the individual would like. The required threshold for trust, though, depends on the importance of the issue. For relatively unimportant issues, one may be willing to seek influence from someone whose intentions are unclear or who may not be as competent as one would desire. With increasing importance, the threshold becomes correspondingly higher, particularly when the desired influence concerns decisions and judgments rather than factual information. In sum, whether a target has sufficient trust in a source reflects an interaction between the importance of an issue and the trustworthiness of the source.

Trust is recognized as a crucial concept in the social sciences, including psychology (Deutsch 1962; Worchel 1979), economics (Berg et al. 1995; Chiles and McMackin 1996), sociology (Gambetta 1988), anthropology (Ekeh 1974), political science (Fukuyama 1995; Barber 1983), and organizational science (Kramer and Tyler 1996). The importance of trust is hardly surprising—no social system could function effectively without some degree of trust among members of the system. In effect, one could look upon trust as the *lubricant* that keeps the motor (i.e., the society) running (Arrow 1974). Without trust, there would be no trade, no social institutions, and most importantly, no friendship, and no love. In more concrete terms, trust has several demonstrable consequences for the functioning of a social system: it makes social life predictable, it creates a sense of community, and it makes it easier for people to work together (Misztal 1998). Trust fosters collaboration and cooperative behavior (Dodgson 1993; Gambetta 1988; Zucker et al. 1996), facilitates conflict resolution (Parks et al. 1996), and increases civic participation (Deluga 1995; Konovsky and Pugh 1994). Trust is critical for the creation and maintenance of social capital (Bourdieu 1997; Coleman 2000; Fukuyama 1995; Zabłocka et al. 2017). As such, trust supports social development (Putnam 1993) as well as the cohesion of a society (Tyler 2003; Cook et al. 2005).

Despite the centrality of trust in theory and research on societal functioning, it is open to different conceptual and operational definitions (cf. Earle 2010; Li 2007; McEvily and Tortoriello 2011; PytlikZillig et al. 2016). There is disagreement, first of all, whether trust and trust exist on a single dimension (e.g., Barber 1983; Deutsch 1958; Rotter 1971; Worchel 1979) or instead function as independent dimensions, with different combinations of trust and distrust possible (e.g., Dimoka 2010; Lewicki et al. 1998; McKnight and Chervany 2001; Sitkin and Roth 1993).

In support of the independent dimensions perspective, positive trust and negative trust (distrust) have different factor structures and different antecedents and consequences (Dimoka 2010; McKnight and Chervany 2001; McKnight et al. 2004; Sitkin and Roth 1993). Also, trust and distrust are associated with different emotional states. Whereas trust promotes a relatively sanguine ("cool and collected") emotional state, distrust promotes a more emotionally-charged emotional state (McKnight and Chervany 2001). In a neuroimaging study (Dimoka 2010), for example, distrust was shown to activate the amygdala, a brain region responsible for intense and sudden emotional reactions. In effect, the speed and intensity of the distrust reaction are due to the role of distrust in safeguarding the person from the potential misconduct of the person he or she is trusting (p. 389).

Distrust thus functions as "the emotion-charged human survival instinct" (Dimoka 2010, p. 378). In this sense, the faster distrust reaction eliminates suspicious candidates from the pool of potential sources of social impact. Trust, meanwhile, signals safety and thus lends itself to cooperation rather than suspicion. Second, some theorists and researchers focus on different aspects of trust, such as the attributes of a trusting person (the *trustor*), the attributes of the object of trust (the *trustee*), the history of their relationship, as well as the attributes of the situational context (PytlikZillig et al. 2016; Li 2007).

The Functional Perspective on Trust We propose that the multiplicity and complexity of trust definitions can be reduced by adopting a functional perspective on trust. In the functional perspective, trust is defined by the role it plays in the life of individuals. By trusting well-informed and dependable others, people can minimize their risks and maximize their outcomes, while making efficient use of their cognitive capacities. From a societal perspective, trust regulates the flow of social influence among people in a society, thus enabling an optimal delegation of information processing.

Trust as Social Judgment Having trust in a source of influence entails social judgment. The research on social judgment is extensive and has identified several important dimensions (e.g., Asch 1952; Fiske et al. 2007; Fiske and Taylor 1991; Wegner and Vallacher 1977). Among the general factors that promote a positive judgment are personal characteristics such as warmth, morality, and conscientiousness, as well as similarity with respect to personality, social identity, attitudes and values, demographics, and social networks. With respect to judgments of trust in particular, several have been identified as critical, including benevolence (Curral and Epstein 2003; Landrum et al. 2013; Schoorman et al. 2007), morality (Wojciszke 2005; Wojciszke et al. 1998), reliability (Johnson-George and Swap 1982), similarity to the trustor (DeBruine 2002; Hayashi and Kryssanov 2013; Ziegler 2013), competence (Curral and Epstein 2003; Mayer et al. 1995; Reeder and Brewer 1979; Wojciszke 2005; Wojciszke and Baryła 2006), and resources (e.g., available time) of the person.

All these factors can be reframed in terms of a sequence of progressively finer decisions. The initial judgment is whether or not the source of influence is viewed as a potential danger. If the source is seen as a potential danger, a judgment of distrust results and terminates any further considerations. This stage is based on the availability of negative information about the source, such as his or her bad intentions, vested interests, undesirable personality characteristics, and questionable reputation.

If the source is deemed not to represent a danger, the next stage of judgment is whether the source is willing to provide help in the processing of relevant information. This stage is based on the availability of positive information, as identified in the literature on social judgment, described above.

If the willingness criterion is met, the next judgment centers on whether the source is capable of providing assistance. The attribute of competence has traditionally been identified as pivotal in making this judgment. Many specific features of competence have been identified and investigated, including efficacy, skill, creativity, confidence, and intelligence (Cuddy et al. 2008). However, other considerations are relevant as well. For example, does the source have the time and energy to provide assistance? Does he or she have the freedom to help? Passing all three criteria promotes a judgment of trust.

In summary, we propose that trust as social judgment functions as an aid for decision making in situations of uncertainty and risk. The decision concerns whether a trustee is a reliable source of information and advice and whether any important aspects of the decision can be delegated to him or her. In the presence of clear negative information, a candidate is immediately rejected; if the information is ambivalent or positive, the trustor decides on the extent and scope of the expected impact based on the assessment of intentions and capabilities of the candidate.

Dynamics of Formulating Trust Judgments: Theoretical Model Most theory and research on trust have treated this construct in static terms, which can be measured at a single point in time, as noted by Rousseau et al. (1998). However, Rousseau et al. (1998) note that trust is a dynamic property that can change over time, at least in organizations. Trust can undoubtedly be stable over time, but it can also vary over time, based on changes in external conditions, such as the behavior of a trustee, the context of decision-making, as well as internal dynamics of the trustor. Viewing trust in terms of *attractor dynamics* enables one to capture both the static and dynamic features of trust (Roszczyńska-Kursińska and Kacprzyk-Murawska 2013). An attractor is a state towards which a system evolves under a given set of parameters or constraints (Vallacher and Nowak 1994) and to which it returns once it has been perturbed. A system can have two or more attractors, each representing a set of states to which the system can converge over time.

With this in mind, trust can be conceptualized as a dynamical system with two attractors, representing trust and distrust, respectively, as depicted in the figure below (Fig. 1.1).

In this landscape metaphor, the movement of the ball is governed by *energy minimization*, such that it will roll to the lowest point in its local neighborhood, in much the same way that a ball will roll into the nearest valley. The bottom of the valley represents the attractor. The shape of the landscape determines the dynamics of the system. The deeper an attractor, the more energy that is required to move the ball out of the attractor. The position of the ball represents a person's momentary level

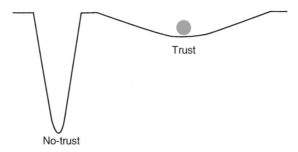

Trust

No-trust

Fig. 1.1 The figure depicts exemplary attractors of 'trust' and 'no-trust'. The state of a person in a given moment is indicated by the *dot*. For a time the person is in the 'trust' attractor. This attractor is broad but shallow, meaning that in most situations the person would prefer to trust; however, this state is very sensitive to external and internal factors. It is relatively easy to push the person to the 'no-trust' attractor which is very deep and it would be relatively hard to push the person out of it

of trust. As depicted in the figure, the attractor of trust has a wide but shallow shape, whereas the attractor for distrust is narrow but deep. A wide attractor for trust means that in a wide variety of situations, people assume trust. However, the shallowness of the trust attractor signifies that the system can easily leave this state and converge on distrust. Once in the distrust attractor, the system is more stable and resistant to change, even when considerable contradictory information is encountered.

This asymmetry between trust and distrust follows from the notion of positive/ negative asymmetry in social judgment (e.g., Peeters and Czapiński 1990). "Positivity bias means that there is a human tendency to generate evaluatively positive hypotheses about reality and a readiness to generate positive affective states … In contrast, the negativity effect is a reaction to specific stimuli and means HIGHER IMPACT of negative than of positive stimuli of the same intensity on behavior" (p. 426, Lewicka et al. 1992). By default, people have a tendency to assume positive expectations and hypotheses about reality, but the impact of negative information highly outweighs the impact of positive information.

In sum, both trust and distrust tend to be self-sustaining in that information inconsistent with these attractor states is unlikely to undermine these states. However, it takes less contradictory information to undermine trust than it does to undermine distrust. People's resistance to contradictory information is, of course, well-documented in research on attitudes and attitude change. Even when people are informed that the information used to generate an attitude is invalid, they nonetheless cling to the discredited attitude (Ross et al. 1975). However, change is possible if the perturbing information is sufficiently strong to bypass a person's psychological defense mechanisms and in effect push the system out of its current attractor and move the system to the alternative attractor. The different shapes of the trust and distrust attractors signify that the transition from trust to distrust is likely to be abrupt and catastrophic, whereas the transition from distrust to trust is more likely to be slow and gradual. This asymmetry is in line with research showing distrust is more emotionally charged and characterized by more sudden onsets (e.g., Curral and Ipken 2006; Dimoka 2010; McKnight and Chervany 2001).

Factors That Shape Judgments of Trust and Distrust Because trust versus distrust are social judgments, the factors identified in the enormous literature on social judgment are relevant to an assessment of someone's trustworthiness. One critical factor is the history of interaction with the trustee. Obviously, the longer the relationship and hence the greater the amount of relevant information concerning the person's trustworthiness, the higher the stability of one's assessment of someone's trust and distrust (e.g., Lewicki and Bunker 1996).

However, people can form trust and distrust assessments of people they do not know well, even of total strangers in the first moments of interaction. Such assessments may be based on diverse information (Mason et al. 2004; Voci 2010). Physical appearance can be determinative, for example, as can nonverbal behaviors, ranging from bodily gestures and posture to eye contact.

1.7.4 Coherence

A third critical factor is the *coherence* characterizing the message, the message context, and the relation between the message and other sources of information (e.g., other available information, existing knowledge). In formal terms, a set of cognitive elements is coherent to the extent that they are internally consistent and consistent in their implications for making a decision or rendering a judgment. A logical contradiction—for example, if A and B co-occur, but A implies not B—exemplifies incoherence. By the same token, if a message both implies that a course of action is sound and not-sound, the message is incoherent.

Theory and research over several decades in psychology, however, paint a far less straightforward picture of coherence. Cognitive dissonance theory (Festinger 1962), for example, has documented the tendency of people to engage in a variety of means to minimize incoherence between their thoughts and actions, or to change the meaning of critical cognitive elements so as to eliminate incoherence. So, for example, a person who views him or herself as socially sensitive but who derogates an otherwise nice person may change his or her view of the "victim" from nice to deplorable in order to eliminate the inconsistency between his or her self-concept and his or her insensitive behavior. If cognitive elements run the risk of appearing incoherent, in other words, people have a tendency to alter the meaning of elements as necessary in order to avoid incoherence. In this view, coherence is the goal of judgment, not the driving force in reaching a judgment. The malleability of cognitive elements points to the inherently subjective nature of coherence as opposed to the a priori and formal nature of coherence.

Despite the subjective nature of coherence, objective incoherence can be noticed and guide information processing and dictate whether or not social influence is accepted. The role of objective coherence versus incoherence has this function when there is an absence of pre-existing information such that trust has not been established and people have not yet made a decision about the acceptance of influence. Some types of incoherence are more likely to be noticed than others. Craig and Lockhart (1972) have shown that information can be processed with varying degrees of "depth." In a written text, for example, first the letters need be recognized, then visual information needs to be converted to auditory code, then the semantic nature of the sentence may be processed to give it meaning.

However, the processing at the semantic level also may be less or more elaborated, either involving just the basic meaning of the main verbs and nouns, or it may also involve a deeper understanding of the relationships between the concepts, elaborating on the implantations of newly acquired information and relation to already existing information. Incoherence is more easily detected at shallower levels of processing, especially if it interferes with processing the information (e.g., grammatical errors) than is incoherence at deeper levels. In fact, because the information may not be processed at deeper levels, incoherence at these levels is likely to go unnoticed.

The perception of incoherence, which is likely when people have little experience with, and trust in the influence agent, can stall the natural tendency to integrate basic information into a higher-order structure with a stable meaning (e.g., Simon and Holyoak 2002; Thagard 1989, 2002; Vallacher and Nowak 1999; Nowak et al. 2000). A set of thoughts relevant to a social judgment is coherent if they collectively convey an evaluation; a set of low-level understandings is coherent if they are sufficiently coordinated to promote effective evaluation. In both cases, the challenge of attaining higher-order coherence may stall the integrative process at a level well beneath that of global evaluation or action mastery. Instead, social judgment will engage a lower level, reflecting a differentiated view by the target (Kunda and Thagard 1996), and understanding will be limited to lower-level facts (Vallacher and Wegner 1985). In line with this reasoning, incoherence limits the individual's vulnerability to persuasion. The websites that are incoherent, for example, are likely to be less effective in persuasion than those which do not disturb the integration process (Tang et al. 2014). "Fake news" on internet sites, for example, propagate more quickly than real news (Vosoughi et al. 2018). Because Fake News contains simple messages and repetitions (Horne and Adali 2017), they tend to create the appearance of coherence, which prevents checking and allows people to conserve their mental resources (Pennycook and Rand 2019). Real news, in contrast, is constrained by the messy, ambiguous, and often equivocal nature of reality, and thus require elaborative processing that strains mental resources.

With respect to trust and distrust judgments, in particular, coherence has two components: the internal coherence of the information per se and the coherence between that information and the information that the target already has. In both cases, incoherence can interrupt the automatic integration process, promoting instead a greater focus on checking the factual information and the conclusions (decisions, judgments, recommendations) provided by the influence agent. The detection of incoherence also undermines trust in the influence agent.

1.7.5 Own Expertise

A fourth critical factor is the target's *expertise* regarding an issue. If the individual has high expertise, he or she is less likely to seek out or rely on a source of influence. This is especially true if the individual has all the relevant information. However, if the issue is relatively unimportant, the individual may still delegate decisions and judgments to others, because this still has the benefit of conserving his or her resources. Moreover, because an expert individual is capable of making sound decisions, he or she is more likely to delegate information gathering rather than decision making to others. An expert individual is also better able to assess information quality and detect incoherence.

Own expertise is also an essential factor deciding about the level at which the influence occurs and also determines the magnitude of reliance on the opinions of others. Even if an individual would prefer to process the information oneself, the lack of expertise may render ineffective the efforts to formulate the judgment oneself and force the individual to relay on the opinion of others.

It is worth noting that expertise is not a stable state but rather can develop over time. A growth in expertise may be driven by other factors, such as importance, trust, and coherence. Because of different combinations of these factors, individuals may not want to delegate information processing, preferring instead to increase their expertise. One possible way for the individual to increase his or her expertise is to delegate both the information gathering and processing to other people and then ask these people to reveal the details of their information processing (e.g., which information is most important, what criteria are used for judgment, what rules are used for integrations, etc.).

1.7.6 Interactions Among the Primary Factors

With the primary factors in hand, we are in a position to illustrate how they interact in the context of the delegation of information processing.

The coherence of information is a primary criterion for trust. In science, the internal consistency of a theory or coherence of empirical evidence is the primary criterion for trusting a theory. In the legal system, the internal coherence of a witness testimony (Glöckner and Engel 2013), agreement between the testimony and external evidence (Uviller 1993), or consistency with acquired narratives (Pennington and Hastie 1992) represent the main factors that jurors use to decide about a witness credibility. It is also common sense that if someone contradicts himself or herself, the message coming from this person raises doubt, which is likely to undermine the person's credibility.

Coherence based on rational considerations may be the primary criterion for trust in science and criminal justice, as noted above, but coherence may be far less rational in everyday life. We propose that in general, any type of coherence supports the primary assumption of trust. Some perceptual configurations are easier to process than others because the various features of the object support one another's interpretations, making for ease of perception, which is referred to as *perceptual fluency* (Johnston et al. 1985; Reber et al. 1998). In the perception of faces, for example, perceptual fluency—and hence coherence—is manifest as symmetry in facial features (e.g., ears, eyes) or as congruence in emotion displayed in different parts of the face. Winkielman et al. (2015), for example, demonstrated that people with faces containing conflicting emotional elements (e.g., smiling lips but a furrowed brow) are trusted less than are faces that consistently express emotions, although this effect is only observed when people are explicitly focused the recognition of emotion. These results confirm the link between coherence and trust and also demonstrate that incoherence needs to be processed to affect trust judgments.

The relation between incoherence and trust is bi-directional. The information from a highly-trusted source is assumed to be coherent and therefore is unlikely to be checked for incoherence. Information from a less trustworthy source, on the other hand, has a higher threshold for coherence and thus is more likely to be checked for incoherence. A highly-trusted politician, for example, can contradict him or herself without producing skepticism on the part of people. However, the slightest evidence of incoherence on the part of a distrusted politician will disrupt the potential for influence from the politician. Even for a highly-trusted politician, however, once incoherence reaches a certain threshold, trust in him or her may decrease, which in turn makes further detection of incoherence more likely. By the same token, if the trust is called into question (e.g., by a moral scandal), previously undetected incoherence in the source's statements may be noticed and become salient and thereby undermine the influence by the source.

Bi-directionality of coherence and trust is consistent with the theory of lay epistemology (Kruglanski and Freund 1983). This theory holds that once people form a judgment about another person (e.g., "I trust him/her")—a tendency called "epistemic freezing"—they become less aware of inconsistent bits of evidence that might force them to revisit their decision/experience/situation. "Unfreezing" is always the option but it is effortful and thus requires motivation to do so.

Coherence and trust interact to determine the level of information an individual will accept from an influence agent. Incoherence prompts individuals to accept only relatively basic, low-level information, reserving the processing of such information to themselves. In fact, incoherence can make an individual wary of even very basic information, inducing him or her to check the validity and reliability of "raw data."

Very high values of trust in the social system, however, prevent individuals from checking information. Under high trust, therefore, inconsistencies are likely to build up and remain unnoticed because low-level information is not processed. Even if information from trusted sources is objectively incoherent, it is compartmentalized so that different items of information are not compared to each other. This effect is characteristic of the authoritarian personality (Adorno et al. 1950; Rokeach 1956). When incoherence is compartmentalized, it may become noticed when there is a sudden decrease in trust. This can produce a feedback loop, in that discovered incoherence further decreases trust and induces low-level information processing, which is likely to promote sensitivity to even higher levels of incoherence. This positive feedback loop may result in a catastrophic collapse of trust. In contrast, the increase in trust is likely to be more incremental.

The general message is that high coherence leads to trust. Low coherence leads to a collapse of trust, checking, and reliance on one's own opinions.

In important matters, when the decision requires information and processing skills that an individual does not have, one possible solution is to ask for advice from someone who is very highly trusted. Alternatively, the individual may ask several experts for advice, and follow their recommendations if they are coherent (i.e., in agreement). If their recommendations and opinions differ, however, the individual is likely to have doubts about their recommendations. Low coherence thus may result in a decrement of trust. When experts are incoherent, the individual may seek out yet more experts and hope that they will provide coherence. Alternatively, the individual may attempt to develop his or her expertise. If this proves successful, the individual may then seek experts' basic information and arrive at a decision based on these facts using his or her own expertise. The route of developing own expertise is the most costly and is likely to be adopted only concerning matters that are very important.

Financial pyramids provide an example of how relying on the opinions of others may be the source of serious mistakes in judgments and decisions. In financial pyramids, trust and coherence play a crucial role. In the case of Bernie Madoff, investors agreed in their perception of the high value of the investment scheme he advocated. It was important that some of the investors were financial experts. High agreement among investors led to high trust, which resulted in it even higher perceived coherence in the opinions of the investors. The effect of this positive feedback loop was that investors relied on the opinions of each other, so that no one was checking the viability of the investment. Once, however, suspicious facts were revealed concerning the investment, there was a rapid change in trust and opinions, in line with the scenario of catastrophe theory (Thom 1975). As individuals started checking the facts, increasing incoherence was revealed. This resulted in even more checking, the results of which further eroded trust.

Interestingly, after the Madoff pyramid was exposed, several other financial pyramids were discovered. The discovery of Madoff's fraud has undermined to some extent general trust in the area of investment. This, in turn, has promoted checking other investment schemes, which in some cases has revealed facts that were not coherent with the narrative of the scheme. Moreover, this has resulted in further checking, often resulting in the discovery of fraud.

1.8   The RTSI Model

The above observations can be formalized as a dynamical model of social influence from the perspective of the target. The RTSI model describes a dynamical system in which social influence is a function of coherence, trust, importance, and expertise of the target. It describes the readiness to be influenced by a source and the type of influence that is sought.

The continuum from information to evaluation (bare facts vs. judgments) is akin to the identification hierarchy proposed by action identification theory (Vallacher and Wegner 1987). It assumes that individuals can process the information concerning their actions at different levels from low-level details (e.g., talking) to the high level (e.g., presenting a lecture, or taking part in the education of the young generation). Higher level identifications are formed in the process of progressive integration of lower-level act identities (Vallacher and Wegner 1987). Following these assumptions, action identification theory states that influence attempts are successful only when information is presented in terms of lower level details. Influence attempts involving higher-level judgments invoke reactance and are resisted, and can result in a boomerang effect – change in the direction opposite to the influence. This conclusion, however, was reached on the basis of experiments that mostly used influence sources that were not trusted – mainly strangers.

Following action identification theory (Vallacher and Wegner 1987), we start from the observation that information and its evaluation form a continuum from low to high level rather than two separate processes. The low level usually corresponds to facts, whereas the high level corresponds to conclusions. Individuals seek to be influenced at different levels of this continuum, sometimes looking for information, at other times for conclusions. The importance of the issue, trust, coherence, and own expertise on the topic are the crucial variables that decide about readiness to be influenced and the level at which the individual seeks the influence. The interaction of these four factors determines the choice of the level. Readiness to be influenced can be recast as a delegation of information collection or processing. The effects of the main variables are described below:

- Importance decides about the level at which individuals seek to be influenced. Individuals tend to be open to information at the high level (conclusions and evaluations) in matters that are of low importance to them, whereas they are open to low-level information (e.g., facts) in matters of high importance.
- Trust influences both the general tendency to be influenced and the level at which the influence is desired. Individuals tend to seek and accept influence from trusted others. Low trust induces low-level information seeking; high trust prompts seeking high-level information. In other words, individuals are ready to delegate the processing of high-level information only to trusted others and will accept the resulting conclusions and evaluations. Less trusted individuals are used as sources of low-level information.
- The more important the issue is to a person, the higher is the level of trust they require in order to delegate its information processing and to be willing to accept high-level influence (e.g., conclusions and evaluations).
- The coherence of messages moderates the relationship between trust and influence level. Checking the coherence of messages provides a primary way for validating messages. Perceived incoherence induces a search for low-level information, even if the information comes from trusted others. Under conditions of very high trust, however, individuals do not check the information for coherence, so even high levels of incoherence are likely to go undetected.
- As a consequence, when trust and latent incoherence are both high, even momentary lowering of trust is likely to result in catastrophic changes, since suddenly incoherence is detected by the individual, which prompts seeking low-level information (e.g., confronting facts) and is likely to lead to the discovery of even more incoherence, further damaging trust.
- Perception of high coherence, in contrast, prompts individuals to accept high-level information and influence even from others who are not highly trusted, which leads to a gradual increase in perception of trust.
- Own expertise influences the tendency to delegate low–level information processing and reserve the high level of information processing to oneself.
- If the initial expertise of the subject is low, the importance of the decisions is high, and the coherence of high-level obtained information is low, individuals are likely to engage in the development of relevant expertise in the area in question.

The system described by these rules has optimizing properties on the individual level, allowing the individual to devote more resources to important judgments and decision tasks. This delegation of information gathering, processing, and decision making to more qualified individuals results in higher quality decisions and judgments.

On the group level, it assembles group members into a distributed information processing system.

1.9 Group Level Dynamical Consequences of the Model

The mechanisms of social influence on the individual level described above dynamically organize the members of a social system into an emergent optimized distributed processing system. Specifically, the rules of social influence decide, based on issue importance, trust, and message coherence, whether an individual seeks low-level information or delegates the processing to others, allowing the individual to accept their conclusions and evaluations. These rules enable information processing resources of the group to be devoted mainly to the processing of important information, delegating the processing to the most capable individuals.

The notion of distributed processing builds on the notion of *transactive memory* (Wegner 1987) by which groups and societies encode, accumulate, and retrieve information and knowledge. TM shows how groups, from those that are small to those that are large (e.g., organizations, societies), develop an elaborate memory system that is likely to be more effective than that of any of the individuals. This system relies not only on information and knowledge but also on the ability of the individuals to access this knowledge (Mitchell and Nicholas 2006). By knowing "who knows what" and "who is the expert," group members can effectively access specific information stored within the group (Yoo and Kanawattanachai 2001). TM provides valuable insights into how specialization benefits a social system. In line with this idea, RTSI specifies that information and evaluation acquired from specific, trusted individuals optimize the functioning of the individuals as well as the whole group.

However, RTSI goes beyond describing the distribution of memory in a group to characterize the allocation of processing to members of the group. Such allocation, in RTSI, is governed by specific rules of influence that represent the interaction of trust, importance, and each individual's level of expertise. The model also reflects a dynamical perspective in which patterns of trust and coherence change by virtue of their interaction in a feedback loop. The dynamic interplay of the model's variables is what enables a group to optimize its functioning.

From the optimality perspective, assigning processing to highly trusted individuals saves the processing resources of the group and allows the distribution of group efforts for specific tasks, where each task is performed by individuals who are most expert and most devoted to processing this information. The theory suggests that there is an optimal level of trust among individuals in a group, with poor group performance associated with levels of trust that are either too low or too high. When trust is too low, the delegation of information processing is hindered, because every group member checks the information. This produces inefficiency because members who have relevant experience not being put to the best advantage. When trust is too high, meanwhile, the group members rely on the opinions and decisions of others, even when checking would optimize the process. In this scenario, the group opinion may be at odds with objective considerations and realities, in a manner that captures the essence of groupthink (Janis 1982) and group polarization (Meyers and Lamm 1976). The optimal level of trust varies in accordance with the evolution of the interactions among the group members' competence, their intentions, the importance of the issue, and the ability of each individual to process information on his or her own. These dynamic relationship among variables are described by RTSI.

Contemporary research in economic and marketing recognizes that the delegation of information processing provides for efficiency and promotes better decisions. Indeed, without delegation of tasks within a group or a social network, individuals would be overwhelmed by the complexity and sheer volume of decisions they need to make in the modern world. Only a small proportion of individuals rationally process the information, with the other group members simply conforming to their choices (Bentley et al. 2011; Simon 2000). Dynamical computer models of this process are used to design marketing strategies and envision strategies for economic growth. In this context, RTSI can precisely specify under which conditions individuals should process information rather than copy the choices of others.

1.10 Summing Up and Looking Ahead

In sum, RTSI makes the following assumptions:

- Social influence is a regulatory mechanism that enables individuals and groups to reduce their action and decision making costs without compromising efficiency, thus optimizing their functioning.
- The optimization happens through active search for influence at different levels of the information and evaluation, with raw facts (low-level information) defining one end of this continuum and opinions and decisions (high-level conclusions) defining the other end of the continuum.
- Social influence processes enable distributed social processing, so that a social group as a unit can process information, produces knowledge, and undertake collective actions. Different portions of information are gathered by specific group members and communicated to others. Various individuals process various chunks of information, arriving at opinions, forming attitudes, judgments, and decisions, which may be communicated to others.
- Relational trust and issue importance provide a bottom-up mechanism that promotes the emergence of a division of information processing among different group members. Individuals tend to seek high level, processed information from highly trusted others. In interactions with less trusted others, however, individuals tend to filter out their high-level opinions and rely only on low level, uninterpreted information. Important issues tend to be processed by individuals themselves based on low-level information, while less important ones are delegated to others.

- Coherence (within and across the sources of information) establishes the threshold for trust. High coherence induces individuals to trust other people's judgments and gradually increases trust in the group, while incoherence prompts individuals to verify evaluations and conclusions by processing low-level information.

We thus claim that social influence optimizes the functioning of groups and individuals by enabling distributed social processing in which information at different levels of organization – from raw facts to opinions – is processed by different individuals and sought by others. The process of actively searching for influence is regulated by trust, the importance of issue, and coherence, and moderated by one's expertise.

In the RTSI perspective, processes of social influence connect different levels of social reality. The influence between individuals underlies group-level processes such as reaching consensus and the emergence of group norms. Because of its strong practical implication, social influence has attracted the interest of practitioners in marketing, social and political campaigns, health-related behaviors, and many other fields. A formal description of social influence, which is precise enough to be used in computer simulations of various social processes, can be used by other scientific disciplines for prediction and designing practical interventions.

The next chapter discusses empirical research designed to validate the primary components of RTSI: trust, the link between trust and coherence, and the importance of the issue. The concluding chapter takes the theory to the group level, using computer simulations to analyze social influence as socially distributed information processing. It also examines conditions under which the delegation of processing to others can lead to errors in processing and hence can result in extreme opinion that are unjustified.

Chapter 2
Experimental Verification

2.1 Experiments on Trust

The functional perspective on trust predicts that trust and distrust have different factor structures. If the function of distrust is to discredit a potential trustee, it should have a simple, one-dimensional factor structure. On the other hand, the decision to trust requires a deeper understanding of the conditions under which a person can be considered trustworthy. If the trust dimension requires more complex information processing, it should also have a more complex factor structure. Specifically, trust should reflect the two-dimensional structure associated with social judgment: intentions (warmth) and capacity (competence).

To investigate the respective factor structures of distrust and trust, Kacprzyk-Murawska (2018) created a questionnaire comprising seven facets of trust that are popular in the literature: *benevolence*, *competence*, *morality*, *predictability*, *autonomy*, *responsibility*, and *calculation*. These facets were represented by 67 items (worded in either a positive or negative direction). The goal of the study was to determine the dimensions of trust that are taken into account when judging someone. There were four experimental conditions representing different descriptions of a person. Two *positive* descriptions portrayed a person who is (a) close to the participant but not a family member, and (b) a trusted coworker. The *ambivalent* description portrayed a person that one does not like but respects professionally. The *negative* description portrayed a person one does not like but with whom the person must maintain contact. For each description, participants recalled a person that he or she knew and assessed that person using the 67 items of the *Interpersonal Trust* scale.

A sample of 606 participants was randomly divided into one of the four description conditions (approximately 150 participants per condition). Participants read the description of a person and completed the *Interpersonal Trust* scale. Separate factor

© The Author(s), under exclusive licence to Springer Nature Switzerland AG 2019
A. Nowak et al., *Target in Control*, SpringerBriefs in Complexity,
https://doi.org/10.1007/978-3-030-30622-9_2

analyses (FA) were conducted for each condition, with oblique rotation because of anticipated correlations among the components. The suitability of FA was assessed before analysis.[1]

Similar 3-component factor solutions were obtained for the positive and ambivalent descriptions.[2] The first factor consisted of items that primarily represented benevolence (but also responsibility, morality, and loyalty). The content of the questions loading on this factor suggests that the factor can be interpreted as the likelihood that the person has intentions to help. The second factor consisted of only negatively-word questions, regardless of their underlying facet of trust; this factor thus represented distrust. This factor represents the judgment of bad intentions. The third factor consisted mostly of items either reflecting the ability to help, competence, or constraints against helping (e.g., timing, availability). For the negative description, a qualitatively different, two-dimensional solution was obtained. The first component consisted of positive items regardless of their prior trust facet, while the second one consisted of only distrust items, i.e., items that were negatively worded (recoded). Unlike the results for the positive and ambivalent descriptions, a third factor did not emerge for the negative description.

The results confirm the hypothesized differential factor structure of trust and distrust, suggesting that the decision of whether to trust is qualitatively different from the decision of whether not to trust. The trust judgment supports a more elaborated analysis of other people's trustworthiness dimensions (like willingness and ability to help), while the judgment of distrust reflects a simple assessment of threat linked to negativity. In the presence of clear negative information -a threat potential – a decision not to trust is made, the person's candidacy for a potential source of impact is rejected, and no finer distinctions follow. However, in the absence of clear negative information, more detailed distinctions concerning a person's positivity become salient in the judgment of the person's trustworthiness. The person's intentions, as well as his or her capacity to help, are assessed.

2.2 The Effect of Incoherence on Trust

RTSI suggests that in a situation in which a *trustor* (i.e., a person making a trust judgment) is familiar with just a few pieces of information about a *trustee* (the object of the trust judgment), the detection of an incoherence lowers the trustor's propensity to trust. We verified this hypothesis in two experiments using a *Trust Game* (see description on page 27). In the first experiment, we manipulated the *level*

---

[1] In all four conditions, the overall Kaiser-Meyer-Olkin (KMO) measure was greater than 0.87 (KMO in 1–3 conditions was over 0.9), meaning the data were adequate and suitable for FA. In all 4 conditions, Bartlett's test of sphericity was statistically significant ($p < .0001$), indicating that the data were suitable for factor analysis.

[2] The rotated solution exhibited 'simple structure' (Thurstone 1947) when loadings above 0.55 were taken into consideration

Table 2.1 Trustees' descriptions used in Study 1

Message subject	Coherent	Incoherent
Obese person	An obese person who in the evenings at home enjoys sweets.	An obese person who declares to be on a diet enjoys sweets in the evenings at home.
An employee of a clothing company	A person who works in the Adidas marketing department privately wears Adidas shoes.	A person who works in the Adidas marketing department privately wears Nike shoes.
Anti-globalist	Anti-globalist took part in the advertising of the Teraz Polska emblem (promotion of local products and services).	Anti-globalist took part in McDonald's advertising.

of coherence by a written description of a potential opponent in a game. A description was composed of two pieces of information: a signal (information about a trustee's behavior) and context (information about trustees' past decisions, appearance or beliefs). A coherent description was composed of a signal which was consistent with the context.[3] We used three coherent and three incoherent descriptions (see Table 2.1). Different contexts and signals were employed to ensure that any observed effects could be attributed to variation in the coherence-incoherence dimension and not to more specific reactions triggered by particular messages.

> **Trust Game**
> Trust game was used and popularized by Berg et al. (1995). This game was originally designed for two players – a trustor and a trustee. At the beginning of the game, each player is endowed with some amount of Experimental Coins (EC). The first move belongs to the trustor – he or she can transfer any amount of experimental coins to a trustee. Anything the trustor spends is then tripled. Then the trustee can return to the trustor any amount of Experimental Coins she or he possesses (i.e., initial endowment plus anything the trustor sent). The trustee's move finishes the game. It is called a trust game because once the trustor sends money to a trustee, she or he loses control over the situation – the trustee can return money or keep all for him or herself. The more money the trustor sends to the trustee, the higher the trust.

As expected, the results, presented in Table 2.2, revealed that participants trusted the coherent trustee significantly more (as assessed by Wilcoxon signed-rank tests) than the incoherent trustee.

In the second experiment, (Nowak et al. 2018), we created pictures designed to vary in their relative coherence. The coherent pictures portrayed individuals or groups in settings that were congruent with a person's physical appearance (e.g., a person wearing a suit in a conference room). Incoherent pictures portrayed a person in a set-

---

[3] The relative coherence versus incoherence of these descriptions was validated in pilot studies.

Table 2.2 The median amount of money transferred to a trustee described as coherent versus incoherent (in Polish złoty)

	Coherent	Incoherent	z
Obese person	3	1.5	2.46**
The employee of a clothing company	4	3	3.88***
Anti-globalist	5	2	6.68***

Note: **p < .01, ***p < .001

ting that was incongruent with his or her physical appearance (e.g., a person in a suit walking on a beach). Each picture was rated by a sample of 250 participants, who judged the picture on 7-point scale, anchored by "very incoherent" and "very coherent."

A separate sample of 916 participants was randomly divided into 4 groups, where each group rated the trustworthiness of a person in a subset of photos on a 7-point scale, anchored by "cannot be trusted at all" and "can be highly trusted." Results revealed that coherence in the photos affected trust ($\chi^2(1) = 51.086$, $p < 0.001$), with more coherent pictures resulting in a higher trust.

2.3 Assumption of Coherence

The assumption of coherence can be described as people's inclination to expect others to behave consistently with their convictions, declarations, and past behavior. Even in a situation in which people possess hardly any information about a trustee, they tend to assume that the trustee will behave coherently. Therefore, people often treat lack of information regarding the trustee's coherence as confirmation of his or her coherence. Positive verification of coherence is treated as redundant with what the trustor already assumes about the trustee and what he or she already incorporated in his or her decision process. In that way, the assumption of coherence saves people from a rather arduous task – a constant verification of others' words and testimonies. It optimizes an individual's behavior – but only in cases of issues with relatively minor significance, such as the cost of verifying a trustee's coherence might exceed the cost of making a mistake. In the case of high importance issues, however, the assumption of coherence is risky and might cause serious problems.

In our research, we tested how the assumption of coherence influences trust. We expected that in a situation in which a trustor needs to decide how much he or she can trust a trustee, additional information confirming the trustee's coherence would not increase his or her propensity to trust. A similar level of trust should be obtained in a situation when the consistency of trustee can be confirmed, and when evidence confirming a trustee's consistency is impossible to obtain.

To verify this hypothesis, we ran an experiment similar to the first experiment. This time, however, participants were led to believe that they would play with real people in real-time, although they were actually playing with an experimental con-

Fig. 2.1 Mean amount of money (0–10 units of experimental coins) sent by the trustor depending on the message received from the partner

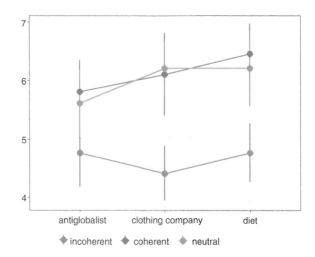

federate. The trust game was played online, with each player located in a separate room. Before the game, players were asked to write one sentence about themselves and send it to the other player via Skype. The confederate used one of three prewritten self-descriptions: coherent, neutral, or incoherent. Neutral self-introduction was constructed from two detached pieces of information that could neither confirm nor challenge the trustee's coherence (e.g., "I am on a diet," and "I like listening to the radio").

The results revealed statistically significant differences in trust as a function of the coherence manipulation. Trustees were given statistically significantly less money after delivering an incoherent self-presentation than after providing a coherent one, $p = .004$, or a neutral one, $p = .01$ (see Fig. 2.1). But, as we expected, there were no statistically significant differences in trust levels toward players who introduced themselves coherently and neutrally, $p > .05$, suggesting that providing additional information that supported trustee's coherence did not increase trust relative to a situation in which a confirmation of coherence was not available (see Fig. 2.1).

2.4   Different Effects of Intrapersonal and Interpersonal Incoherence on Trust

According to RTSI, incoherence affects trust as well as the decision-making process by lowering both the readiness to trust and the probability of accepting the trustee's recommendations. To verify this hypothesis, we conducted two experiments, in which we tested the extent to which the inconsistency can (1) reduce the chances of receiving the influence, (2) decrease willingness to make a decision based on information delivered by the incoherent source, and (3) enhance the probability of searching for additional information. Two types of incoherence were examined: intrapersonal and interpersonal.

We asked participants to read two sets of movie reviews and indicate (1) the general trustworthiness of different film critics, (2) how much they trust the film critics, and (3) their readiness to make a decision based on the reviews. Incoherence was operationalized as inconsistency in movie reviews. We prepared two sets of four movie reviews – a coherent and an incoherent set. The incoherent review was always the last one. In the first experiment with intrapersonal coherence, all the movie reviews in the single set were posted by the same author. In the second experiment with interpersonal coherence, the same set of movie reviews were used, but participants were led to believe that different authors wrote each movie review and posted it on the same online forum.

2.5 The Effect of Incoherence on Trust and Trustworthiness

The results of the experiment with intrapersonal coherence confirm our other results obtained from the experiments on the influence of incoherence on trust. Participants trusted authors of coherent reviews more than authors of incoherent reviews, $t(36) = 5.64$, $p < .001$. As expected, an author of the incoherent review was judged as less trustworthy than the author of incoherent reviews, $t(36) = 4.65$, $p < .001$, and the probability of following his or her recommendations in the future was lower by 20.27% points, $t(36) = 4.34$, $p < .001$.

Similarly, in the experiment with interpersonal coherence, in which each review was written by a different author, an author of the incoherent review was trusted less than authors of coherent reviews, $p < .01$. No differences in trust level toward authors of consistent reviews were found, $p = .58$.

However, the decrease of trust toward the incoherent author did not affect the evaluation of movie critics as a group. When we asked participants to indicate how much they trust the movie critics as the group, they did not report greater trust for the group of coherent authors than for the group of incoherent authors, $p = .85$. Additionally, we found that participants declared similar probabilities of searching for the reviews posted on the online forum with coherent and incoherent movie reviews in the future, $p = .22$.

These results suggest that when the source of incoherence can be unambiguously diagnosed (the incoherent author in case of our experiment), the effect of incoherence affects only the source of incoherence, and it does not generalize to the whole group. The group composed of incoherent authors can be trusted as much as the group whose members have not shown signs of inconsistency.

2.6 The Effect of Incoherence on Readiness to Make a Decision

According to RTSI, the effect of incoherence is not confined to trust. Incoherence lowers confidence in an information source, which in turn diminishes the satisfaction in information sent by the incoherent source and reduces the readiness to make a decision based on an unreliable source. We expected that a decision-maker facing an incoherent source of information should feel compelled to search for additional information from a source that can be trusted. Therefore, we asked participants two questions: (1) Were they were ready to choose the better movie based on available information, and (2) whether they preferred to read additional information about the movie before making their final decision.

In line with our expectations, people were not satisfied with reviews delivered by an incoherent author, and they expressed a desire to read other reviews before choosing the movie, p = .007. After reading coherent reviews, however, participants were more ready to decide on the movie than after reading inconsistent reviews, p = .003.

Quite different results were obtained from experiments in which interpersonal coherence was investigated. The number of undecided participants was not larger when different authors provided different reviews than when different authors provided similar reviews, p = 1. Also, there were no statistically significant differences in the number of participants who declared that they did not feel the need to read other reviews before making up their mind, p = .69.

These results indicate that people deal differently with incoherence depending on its origin and to whom they have access at the given moment. When an incoherent person is their only source of information (i.e., intrapersonal incoherence), they experience difficulties in making a decision. They also tend to want more information before making a decision. However, when incoherence reflects different opinions expressed by people with coherent opinions (i.e., interpersonal incoherence), people's decision making is not affected. People are ready to choose a movie without any further investigation as if they did not experience any incoherence.

2.7 The Effect of Incoherence on the Decision When No Verification of Obtained Information Is Possible

We expected that an incoherence of a trustee would negatively influence the evaluation of a movie. Accordingly, we tested the influence of incoherence on the overall assessment of the film and the probability of watching it.

For intrapersonal incoherence, analysis of participants' movie assessments revealed that more subjects (57%) rated the movie with incoherent reviews more negatively than the movie with coherent reviews, although this effect was only marginally significant, p = .06. Similar results were obtained when participants were asked to assess the probability of watching the movie in the future: 57% of participants indicated a higher likelihood of watching a movie with coherent reviews than movie with incoherent reviews, p < .05.

When the participants were asked to choose the better movie or to pick up the DVD with the movie they liked more, more participants chose the movie with coherent reviews than with incoherent reviews. Both differences were in expected direction, although chi-square goodness-of-fit tests showed that differences were not significantly different when choosing the better movie, $\chi^2(1) = 2.78$, p = .1, and marginally significant when choosing the DVD, $\chi^2(1) = 3.27$, p = .07 in case of choosing the DVD.

For interpersonal incoherence, there were no significant differences in participants' assessment of the movie with coherent and incoherent reviews, z = −.179, p = .87. Moreover, there were no differences in probabilities of watching the movie with coherent and incoherent reviews, p = .44. There were no statistically significant differences in the number of participants choosing the film with coherent reviews compared to a film with incoherent reviews, p = .87.

In summary, we found that incoherence affects the final assessment of the movie, but only if the incoherence appears to be intrapersonal. When incoherence is created interpersonally (there is disagreement among two or more sources of information), the movie with purely coherent reviews does not have a favorable advantage.

2.8 The Influence of the Expertise of Information Source on the Perception of Coherence

RTSI holds that there is a bidirectional relationship between coherence and trust. The experiments described above demonstrated that incoherence undermines trust. However, it is also the case according to RTSI that the judgment of trust decreases the tendency to check for coherence, so incoherence tends not to be detected in highly trusted individuals. This hypothesis was tested in the following experiment.

We expected that the performance of non-specialists would be assessed as more incoherent than the performance of a professional.

We conducted an experiment investigating the influence of expertise on the perception of coherence. Using an online platform, we asked 223 participants to read a one-page long text and to assess its coherence. We created two versions of the text: consistent and inconsistent. In the inconsistent text, mistakes concerning vocabulary, style, and grammar were embedded. The text was introduced to participants as a part of a novel written by an apprising author or by the already well-acclaimed writer. Of the original 223 participants, we analyzed data from only 88 participants who spent at least 2 min reading the text. Results revealed a statistically significant interaction between the consistency of the text and the expertise of the writer on the perception of coherence, p = .033. Specifically, the consistent text was evaluated as more coherent than inconsistent text, but only when it was presented as the work of an acclaimed writer, p = .001. When the consistent text was presented as a piece of writing of an apprising author, it was evaluated just as incoherent as an inconsistent text (Fig. 2.2).

Fig. 2.2 Mean values of perceived incoherence (scale from 1 to 5) as a function of the consistency of a text and the level of the writer's reputation

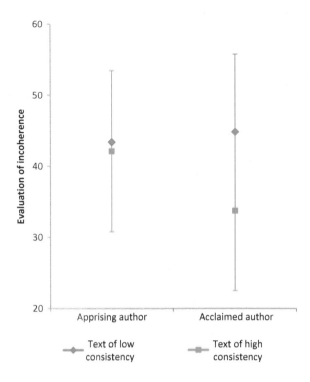

2.9   The Assessment of Incoherence

In another experiment, we investigated what makes a piece of text incoherent? Is it grammar, semantics, or both?

We asked 502 participants to read a short fragment of a text and then evaluate its coherence using a questionnaire designed to measure the text's integrity. We created five different versions of the text. The first (default) version was a fragment of a well-written novel. Other versions were modifications of the default text. The modification was created by introducing different three forms of inconsistency into the text: *semantic inaccuracy* (second text), *grammar errors* (third text), and *both semantic and grammar mistakes* (fourth text). The fifth version was the same as the fourth version but with the first sentence written correctly. This was done to examine whether a coherent beginning of a text can lull the reader into "automatic" mode in which it is easy to overlook signals of incoherence. Each participant received just one text for evaluation.

Results revealed statistically significant differences in mean evaluation of coherence between different texts, $p < .001$ (see Fig. 2.1). Participants judged texts with grammar mistakes, and texts with both semantic and grammar mistakes as less coherent than default text. When the first sentence was correct (in the 5th version),

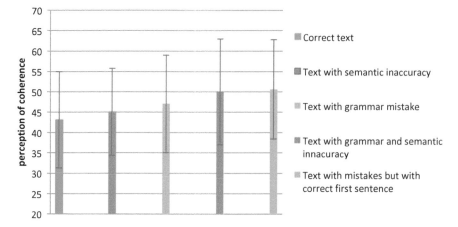

Fig. 2.3 The mean assessment of coherence of text versions differing in their forms of incoherence

the semantically and grammatically inconsistent texts were still evaluated as more incoherent than the default text. Surprisingly, the semantically inconsistent text was not evaluated as more inherent than was the default text, p = .28. These results suggest that people screen text for incoherence based primarily on grammatical features. Inconsistencies at the semantic level do not get detected automatically. Shallow processing uses only a small amount of processing resources. Deeper processing, necessary to detect semantic incoherence, is not performed unless individuals enter a *check mode* on the basis of other available information (Fig. 2.3).

2.10   Importance of the Issue and Trust Toward the Source

Bui-Wrzosinska et al. (2016 unpublished manuscript) conducted two experiments that investigated how the level of information sought from an influence source is determined by the main and interactive effects of trust in the source and the importance of the issue.

2.10.1   Experiment 1

In the first experiment we manipulated trust as a between-subjects factor and importance as a within-subjects factor. Using eye-tracking equipment, we measured the amount of time participants fixated on high-level information (conclusions, opinions) and low-level information (opinion-relevant facts).

Participants   The participants were students of SWPS University in Warsaw, who were recruited by in exchange for credit for coffee in the University's canteen. Participants were informed that they would take part in a 15-min decision making-experiment with the use of an eye tracker. Participants (N = 124) (93 females, 24 males, 7 who did not specify gender), who ranged in age from 18 to 53 years (M = 26.4, SD = 5.52), were randomly assigned to the experimental conditions.

Procedure and Design   All participants were told that they would be presented with a 10 min. Presentation, during which they would be asked to assess four different products, each depicted with specific information presented on a slide for 15 s. After each slide, participants rated the product on a 5-point scale, with higher numbers indicating higher positive evaluation. In the *trustworthy* condition, we informed participants beforehand that information about products was taken from a popular and trustworthy website. In the *untrustworthy* condition, the source website was described as suspicious and untrustworthy. Each slide contained a photo of the product in the middle, the price of the product at the top, and two columns of information about the product at the bottom on the left and right sides, respectively. One column consisted of facts concerning the product, while the other consisted of opinions regarding the product. The association between information type (facts vs. opinion) and column positioning (left vs. right) was varied randomly across participants and conditions. Controlling for the left versus right positioning of low- versus high-level information was essential because of the left-to-right reading bias in Poland.

The same four products were presented in both the trustworthy and untrustworthy conditions. The importance of each product was operationalized in terms of its price. Products such as a laptop and an SLR camera were priced ten times higher than a digital picture frame and an automatic camera. Because the higher-priced products had a value that approximated the average monthly income in Poland, we assumed that they would be viewed as correspondingly important. Each presentation consisted of two slides with products characterized by high importance and two slides with products characterized by low importance. We varied the temporal order of low- and high-importance slides within each between-subjects condition.

Eye-Tracking   The direction of gaze toward facts or opinions was measured by eye- tracker. Participants were seated in front of a 1700 TFT Tobii T60 monitor. Slides were presented on the monitor using Tobii Studio software. During the presentation, the Tobii monitor recorded gaze location and pupil diameter for both eyes based on the reflection of near-infrared light from the cornea and pupil. Gaze and pupil information were sampled at a frequency of 60 Hz. Monitor specifications included accuracy of $0.5°$ of the visual angle and tolerance of head movements within a range of $44 \times 22 \times 30$ cm.

Missing Data   Due to a programming mistake which affected RT measurement, we had to remove data from one low importance item slide in the untrustworthy condition. We, therefore, removed from further analysis an equivalent slide in the trustworthy condition.

Results We used R^4 to perform a linear mixed effects analysis. The fixed effects were trust (trustworthy vs. untrustworthy source), item importance (high- vs. low-priced), information level (facts vs. opinions), and the respective positioning of the information (left vs. right for facts and opinions). The random effects were participants and questions. The dependent variable was the length of visual observation of facts and opinions (left and right sides).

Because we were primarily interested in the interaction models, p-values were obtained by likelihood ratio tests of the full model with the effects in question against the model without interactions.

The fitted model revealed several significant effects. There was a longer fixation time on the right column than on the left column (33645 s ± .5899 greater, $\chi^2(1) = 38.48, p < 0.0001$), which may be indicative of left to right bias in reading in Polish. Additionally, there as a significant level of information effect, $\chi^2(1) = 4.32$, $p < 0.05$, such that fixation time was longer for opinions than for facts.

Of greater theoretical interest are the interaction effects. There was a significant interaction between level of trust and level of information ($\chi^2(1) = 7.10, p < .01$), such that for the trusted source the participants tended to look both for facts and opinions, but for untrusted sources they looked less at the opinions than the facts. This interaction is illustrated in Fig. 2.4, which plots confidence intervals of marginal means of observation length as a function of the level of information and level of trust. Also, the length of opinion observation was significantly greater ($p < .001$) in the trustworthy than in the untrustworthy condition. Taken together, these results suggest that when provided a trustworthy source, people are interested in both facts and opinions, but that when provided an untrustworthy source, people are focused primarily on facts and tend to disregard opinions. People need high trust to be interested in opinions.

There was also a significant level of information * level of importance interaction, $\chi^2(1) = 7.33, p < .001$. This interaction is illustrated in Fig. 2.5, which plots the confidence intervals of marginal means of observation length as a function of the level of importance and the level of information. The pairwise contrast analysis revealed that facts describing more important products are observed longer than are facts describing less important products ($p < .01$), but observation time of opinions did not differ for important and unimportant products. These results suggest that when a source is important, people attend to facts rather than to opinions, but when a source is unimportant, people focus their attention on opinions somewhat more than on facts.

---

[4] All statistics were computed with R version 3.4.1. The packages "dplyr" (version 0.7.4), "magrittr" (version 1.5), "tidyr" (0.7.2) and "ggplot2" (2.2.1) were used for data manipulation, processing and visualization (Wickham 2009). The R package "lme4" version 1.1.14 was used for the mixed-model analysis reported herein.

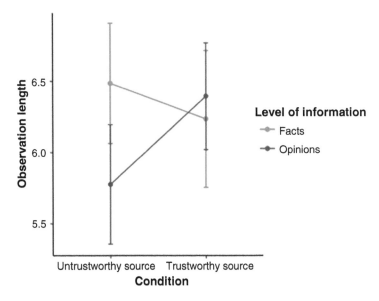

Fig. 2.4   Interaction plot: Levels of trust versus Levels of information

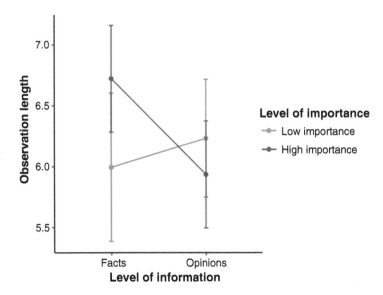

Fig. 2.5   Interaction plot: Levels of information versus Levels of importance

2.10.2 Experiment 2

The goal of this experiment was to investigate how the time needed to evaluate information is determined by the main and interactive effects of trust in the source and the importance of the issue. We investigated these factors using the same experimental design employed in Experiment 1. Unlike Experiment 1, however, we measured the time of decision making instead of the length of fixation on opinions (high-level information) and facts (low-level information). The level of trust was manipulated as a between-subjects variable and the level of importance was manipulated as a within-subjects variable.

Participants The participants (58 women, 31 men) were students of SWPS University in Warsaw, recruited in exchange for credit for coffee in the University's canteen. Participants were informed that they would take part in a 15 min. Decision-making experiment. The participants ranged in age from 18 to 47 (M = 24.7, SD = 5.83), were randomly assigned to each condition.

Procedure and Design We used the same materials as in Experiment 1. However, instead of using an eye-tracker, we measured the time participants needed to evaluate the product.

Results We used R^5 to perform a linear mixed-effects analysis. The fixed effects were the level of trust in the source (high vs. low trust), the importance of the item (high vs. low importance), and the level of information (opinions vs. facts). Intercepts for subjects and questions were random effects. The dependent variable was the time participants needed to evaluate each product. We transformed this measure to the natural logarithm of time in order to normalize residual plots and to maintain homoscedasticity of the variance. p-values were obtained by likelihood ratio tests of the full model with the effect in question against the model without the effect in question.

The fitted model revealed several significant main effects. The level of the information affected the time of decision making ($\chi^2(1) = 15.80, p < .001$). Decision time was longer in the untrustworthy condition, indicating than when trust was low, individuals tended to process the information themselves, rather than rely on the opinions of others. Level of information affected the time of decision making ($\chi^2(1) = 4.08, p < .05$), but only for information displayed on the left side of the slide. Decision time was longer for facts displayed on the left side than for opinions displayed on the same side. The importance of the product affected the time of

---

[5]All statistics were computed with R version 3.4.1. The packages "dplyr" (version 0.7.4), "magrittr" (version 1.5), "tidyr" (0.7.2) and "ggplot2" (2.2.1) were used for data manipulation, processing and visualization (Wickham 2009). The R package "lme4" version 1.1.14 was used for the mixed-model analysis reported herein.

decision-making ($\chi^2(1) = 15.8$, $p < .001$). Decision time was longer for products of high importance.

Taken together, these results suggest that individuals tend to process information themselves what results in longer reaction times when the decision concerns important products and when the source is untrustworthy.

2.11   Summing Up

The results of the experiments we have described support the predictions derived from RTSI. The first study confirmed that judgments of the trust consist of three dimensions corresponding to three primary concerns: (1) Does the person represent a potential danger? (2) Is the person willing to help? (3) Is the person capable of helping? If the first concern is answered positively (i.e., the person may be dangerous), the third concern effectively becomes irrelevant. The judgment of negativity, in other words, stops the processing of further information regarding a source of information.

The second study showed that when there is minimal information concerning the source, incoherence in the source's characteristics undermined trust in the source. The third study generalized these findings to incoherence generally, so that even pictures that portray individuals in unusual contexts undermine trust in these individuals. The fourth study showed that people tend to assume coherence in the absence of information so that when there is no information, people experience the same level of trust as they do when they are presented with coherent information. In contrast, the presence of incoherent information undermines trust.

The fifth study compared the effects of intrapersonal and interpersonal incoherence. The results revealed that when an individual expresses inconsistent opinions, his or her trustworthiness is undermined, and the probability of following his or her recommendations in the future was lowered. The results of the study also showed that intrapersonal coherence of information promotes readiness to make a decision, whereas incoherence promotes a tendency to collect more information. Intrapersonal incoherence also negatively affected the evaluation of the recommended product.

When incoherence concerns one individual who expresses inconsistent opinions, from several others, that are mutually consistent, the trustworthiness of the dissenting individuals was undermined., but the trust toward authors of mutually consistent reviews were not affected by one dissenting individual, nor trust toward the whole group of reviewers. Also probabilities of searching for more reviews posted on the online forum were not affected by the interpersonal incoherence of the movie reviews. These results suggest that when the source of incoherence can be unambiguously diagnosed (the incoherent author in case of our experiment), the effect of incoherence affects only the source of incoherence, and it does not generalize to the whole group. Also then the incoherence concerns one dissenting voice, interpersonal incoherence does not result in the increase of the tendency to search for more information, or in the lower evaluation of the product.

The sixth study showed that incoherence of a less trustworthy (on a competence dimension) source is detected more readily than the incoherence of a trustworthy source. The seventh study has demonstrated that whereas grammatical incoherence is detected automatically, semantic incoherence tends to go undetected.

The sevens and eights studies showed that when trust is low, and the issue is important, individuals tend not to rely on opinions of others, but rather look for facts, and tend to process information themselves. However, when trust is high, and the issue is unimportant, individuals tend to both look for facts and accept opinions, saving their processing resources.

Chapter 3
The Social Group as an Information Processing System

In everyday life, we make dozens of decisions each day – some of them small, some large, some trivial and a few of dire importance. We choose what to buy for dinner, which movie to see, what diet to start, what to wear for a friend's wedding, what car to buy, whether to take a mortgage and sometimes even how to live our lives. Upon these decisions rests our welfare and often also the wellbeing of various social groups to which we belong such as family, friends, coworkers Opinions and choices, therefore, carry a functional value.

Opinions and choices matter. Some are good, and some are bad, for each agent and the social system as a whole. However, many opinion dynamics models concentrate on the "dynamics" but rarely take any interest in the "opinion." They almost unanimously treat opinions as neutral bits of information. In such models, it is important whether any candidate will win, and what the dynamics of the process are, but it is not important what the candidates' stand for. In reality, however, the opinion is what matters. Often the dynamics of the spread of opinion depend on the specific opinion at issue. How do individuals and groups decide which option is good and which is bad? What process allows individuals and groups to reach a useful decision or make a correct choice?

Provided that the situation is within an agent's capabilities to assess, he or she would most often be better off by checking the facts and deciding on his or her own. However, if he or she were to do so on every occasion and for every problem, he or she would have scant time for anything else. Moreover, for some problems, the agent's competence to assess the facts could insufficient. Therefore, an agent relying on his or her own wits in every instance is simply impractical, f not impossible. Fortunately, individuals are surrounded by and connected to many similar individuals who also gather and process information and are often willing to share the results. That is why it is so convenient to take advantage of one's social group's resources – and resign oneself to some degree of informational social influence (Deutsch and Gerard 1955; Festinger 1950).

A. Nowak et al., *Target in Control*, SpringerBriefs in Complexity,
https://doi.org/10.1007/978-3-030-30622-9_3

Yet, the accuracy of the decision an individual makes after consultation with others is only as good as the information they provide. Is the socially agreed-upon assessment by coworkers of the boss's preferences good enough to buy a fitting gift for her birthday? The answer depends in part on whether anyone in the group has had previous experience that are relevant and introduce accurate facts into the discussion. With such facts, the other group members can update their shared reality.

Optimality of individual decisions is therefore tightly linked to the quality of group opinions – that is, to the accuracy of the shared reality of group members (Festinger 1950). This accuracy is established mainly through\individual actions that contribute to the group's knowledge and memory (Wegner 1987). This process is continuously updated and negotiated through deliberation and social influence. Some groups are more successful in this process than are others. For example, the phenomenon of groupthink (which can have dire consequences, as described in Janis 1972) can be described as a failed optimization process, in which updates from outside of the group are not accepted. In more general terms and broader contexts, such a situation is captured by the echo chamber effect (e.g., Uzzi and Dunlap 2005; Colleoni et al. 2014). On the other end of the continuum of possible outcomes, we have a "group" – or rather an assembly of individuals – the members of which are disconnected from each other, and do not rely on social capital in their decisions and actions (e.g., Putnam 2001).

This seemingly natural mode of individual-group interaction can produce pathologies such as a dysfunctional echo-chamber or a collapse of the collective. If group members only communicate unsubstantiated opinions, neither individuals nor the group as a whole will reach accurate decisions. On the other hand, if group members only rely on their wits and never use the opinions received from others, the collection of group members does not truly constitute a group, but rather would simply be a loose assembly of individuals, each working and living on his or her own. What is it that prevents groups from going to these extremes? What are the regulatory mechanisms that, on one hand, enable most individuals to draw from the wisdom of their social context effortlessly but at the same time prevent total free-riding that would lead to a divergence of socially shared reality from actual reality?

In what follows, we present an agent-based model that investigates the group mechanisms that underlie group opinion formation, based on the assumptions on the RTSI. We show that a simple feedback loop between group opinion coherence and trustfulness of individuals may lead to either an echo-chamber or a collapse of social ties. We then expand the model to explore different ways for a social system's assessment (opinion) to flexibly follow the ever-changing reality without any significant loss of group integration. To achieve such a "sweet spot" that enables the group to remain up to date but also efficient in supporting individual and group decisions, the group needs to employ proper social networking mechanisms. What is also crucial is the ability of individual members to set their opinion acceptance range optimally. We show that these mechanisms lead to an emergent diversification of group roles that on the one hand allows for opinion formation in the process of social influence but at the same time prevent the group from straying too far from reality.

3.1 The Model

The model construction and mechanics are derived from the RTSI, combining its pivotal variables and their interactions. By implementing these assumed regularities into the mechanisms of agent behavior and their mutual relations, we test the implications of the theory for group-level dynamics – especially, for optimality of group decision making.

3.1.1 Assumptions

Our main theoretical proposition is that social groups evolved to optimize both individual and group functioning through their ability to make the best use of member competencies through the division of tasks. For this process to work, the group needs to regulate information gathering and processing. We propose that trust serves as a regulatory mechanism for social systems in that it allows individuals to switch between relying on social influence and relying on one's own processing capabilities. We distinguish between different facets of trust: trustfulness, which is a property of each individual and is a generalized propensity to trust specific others; and trustworthiness, which is a property of a directed, trusting relation between any two individuals. We treat both types of trust as dynamic properties that are dependent on the interaction history within a social group. Trustfulness is contextually linked to the perceived reliability of group processing. In particular, we propose that coherence of opinions among group members is an important indicator of such reliability and therefore strongly influences trustfulness. On the other hand, trust relations reflect the subjective, perceived reliability of particular social contacts of an individual and therefore depend on the history of his or her interactions and the properties of the trustees.

Below we describe each of these assumptions in detail, backing each of them with relevant theoretical literature from which it was derived and where applicable, with empirical findings that confirm its role.

3.1.1.1 Social Groups Are Information Processing Systems That Can Optimize Group and Individual Decisions

First of all, a group is more than the sum of its parts in that individuals within a group gain some advantages over those of individuals acting alone. For opinion formation and opinion spread, this assumption means that groups can gather and process information in ways that surpass these actions when performed by individuals. In short, groups, like individuals, are information processing systems.

One of the most important functions of social groups is providing their members with an interpretation of reality and knowledge about what is right and what is wrong (Festinger 1950, 1954; Sherif 1936; Sherif and Sherif 1964; Hardin and Higgins 1996; Bar Tal 2000; Levine and Higgins 2001). This interpretation is constructed through the process of social interaction in which information, such as judgments, categories, and beliefs are deliberated on and distributed between members (Kruglanski et al. 2006; Back 1951; Festinger 1950, 1954; Schachter 1951; Asch 1955; Deutsch and Gerard 1955). Once established within the reference group, people tend to rely on the validity of the socially constructed attitudes and opinions (Festinger 1950).

A group in which interpretations of reality are shared becomes an independent system able to process information. In his theory of distributed cognition, Hutchins (2001, 2006) argues that the cognitive processes that take place at the level of the individual mind can be performed at the level of a social group (Hutchins 2001). According to Hutchins, learning is a process that takes place between people, objects, artifacts, tools, and the environment; the cognitive process is not limited to a single person, as in traditional theories of cognition. A cohesive group can operate as a distributed computing system or an emergent mind. In such a system, the cognitive effort is divided among members who share and negotiate knowledge. In the process of social interaction, collective forms of individual cognitive processes–memory, decision making, the formation of opinions—are performed: (Hutchins 2006).

Wegner (1987) proposed a similar mechanism of collective information processing in his theory of transactive memory. Within a transactive memory system, each member is can estimate who in the group is able to provide information on the relevant topic. Through the process of memory allocation, units of the system become increasingly specialized and shaped in a manner that is most optimal for the group. Specialization of knowledge developed within the transactional memory system extends the overall group expertise and reduces the redundancy of individual cognitive effort (Hollingshead 1998). On the other hand, the complexity of the memory transaction can cause confusion, especially when the allocation of expertise is disputable (Wegner 1987).

3.1.1.2 Trustworthiness and Trust(Fulness) Determine Group Interactions

For a group to effectively serve the role of an information processing system that optimizes opinions and the resultant decisions, there needs to be differentiation of expertise, skills, and roles within its structure. This differentiation is reflected in the actions of advice and opinion seeking, which, with time, tend to be calcified into trustworthiness relations. That is, each individual trusts selected others within each problem domain, based on his or her previous experience with that social contact.

While trust relations in real social systems are often determined by the trustee's characteristics (related to her competence or expertise, or moral qualities), trust is often understood as an individual, psychological trait – a generalized propensity to trust others. Therefore, we assume that trustworthiness relations are separate from the tendency to trust others – that is, trustfulness. We base our assumption on the distinction delineated in the literature on trust, between the trustworthiness of the trustee and trustfulness of the trustor (e.g., Li 2007). Trustworthiness is often linked to a psychological attitude towards a source with a given reputation, while trustfulness is regarded as a choice or a disposition to place trust in others (Dietz and Den Hartog 2006; Lewicki et al. 2006).

3.1.1.3   Trustworthiness Depends on the Reliability of Social Contacts

These properties of the trustor and trustee are not set in stone – they change in time depending on the interactions within the group.

Trustworthiness relations can change due to a variety of mechanisms. First, people tend to trust more those who express similar opinions, even though they necessarily lose access to potentially useful information when it is dissimilar to their opinion. Moreover, people tend to dismiss opinions from others who cannot seem to make up their minds by, for example, changing their assessment too often or too diametrically. Finally, the most potent cue for distrust is access to objective facts that contradict the assessments made by individuals' social links.

3.1.1.4   Trustfulness Depends on the Reliability of the Group as a Whole

Trustfulness may be changed when the group improves or declines in its performance as a source of reliable information. If the group can arrive at a shared representation of reality, its members tend to become more trustful. When there are many disparities and disagreements, the members might become distrustful towards all of their social links. In effect, coherence of opinions might serve as a barometer of the reliability of the group.

People treat other people's opinion as social proof (Cialdini 1984), provided there are other sources that confirm it. Multiple source effect (Harkins and Petty 1981, 1983, 1987) indicate that people believe in information more when they hear it from multiple sources than from a single source. For example, in two studies by Lee and Nass (2004), participants heard five positive reviews of a book from five different synthetic voices or a single synthetic voice. They tended to evaluate the book more favorably when the opinions were stated by multiple synthetic voices, even when they realized that the voices were artificial. The literature on multiple source effects shows that concordance of information from different sources changes disputable viewpoints into "objective facts."

3.1.1.5 What Is Perceived as Coherent Depends on Individual Opinion Acceptance Range

To determine whether the opinions are indeed consistent among group members, one needs to decide what level of differences matters, that is, what constitutes similar (shared) and dissimilar (divergent) opinions. Even if the decisions based on such opinions are binary (e.g., vote for candidate X in the elections), in the process of arriving at them the individual and other group members move back and forth on the continuum between "yes" and "no" to finally decide one way or another. If the decisions and opinions naturally form a continuum (e.g., how good was the blockbuster movie?), the dynamics of wandering between extreme opinions are even more visible. If in this process members of the group rely on social influence and need to assess how coherent the group opinions are (and following that, how much can the group be trusted), they need to determine what range of dissimilarity is acceptable.

The notion of opinion range acceptance has already been used in agent-based modeling, albeit in a different form. In the Weisbuch-Deffuant bounded confidence models, opinions of others are important only when they are within limits of acceptance (Weisbuch et al. 2003). When differences between opinions are large, no "dynamics" (i.e., opinion "adjustment") can occur, as the opinions are not close enough to be "adjusted." Instead, opinion holders restrain from any discussion, as it could quickly turn into conflict or at least a serious argument. Simulations in which opinion exchanges between agents are limited by a proximity threshold (e.g., Hegselmann and Krause 2002; Weisbuch et al. 2003) show that the restraint on discussion promotes clustering and fragmentation of opinions within a group. The lower the threshold, the greater the tendency for distinct groups holding unique opinions to survive, which decreases the likelihood reaching global consensus (Weisbuch et al. 2003).

Bounded confidence is similar to other social psychological concepts concerning opinion acceptance and attitude formation. An important aspect of attitude structure, according to the social judgment theory, is the width of the latitude of acceptance (Eagly and Telaak 1972). This feature of attitude structure predicts how a person will react to novel information concerning the attitude object. If a person has a wide latitude of acceptance on an issue, he or she is likely to change his or her attitude in order to incorporate new information in his or her cognitive system, even if it is highly discrepant from the central tendency of his her own opinion. If there is narrow latitude of acceptance, on the other hand, there is correspondingly little chance that new information will be acknowledged and promote a change in attitude. A parameter of latitude of acceptance thus predicts the magnitude of information discrepancy necessary for the discrepant information to be rejected.

The latitude of acceptance determines individual susceptibility to social influence concerning a given issue. Hovland et al. (1957) observed that subjects with extreme opinions accepted fewer persuasive statements and rejected more statements, which resulted in them having smaller latitudes of acceptance and greater latitudes of rejection than subjects with less extreme positions.

A study by McCroskey and Burgoon (1974) shows that a person's widths of latitudes of acceptance and rejection are relatively stable across topics and they are not related to attitude's polarity. Eagly and Telaak (1972) show that the latitude of acceptance width predicted attitude change better than did discrepancy of the persuasive message itself.

Based on these concepts, we assume that individuals perceive opinions to be divergent if the difference between them is above a certain threshold. This way, each member of the group can assess how coherent the group is as a whole. Next, according to the Regulatory Theory of Social Influence, we assume that incoherence within the group reduces member's propensity to trust others from the group, and coherence increases this propensity (i.e., trustfulness). If trustfulness is high, the individual relies on social influence to arrive at her opinions; if it is low – she invests her own cognitive effort to analyze the situation and make up her mind.

To test how well trust serves the role of regulating information processing in a social system, we have implemented the above assumptions into an agent-based model, as described below.

3.2 Implementation of Model 1

3.2.1 Structure

The multi-agent model is designed to mimic a medium-sized social group that estimates some real-world situations or state of affairs such as collective decision making or initiating collective action. The situation the group assesses could range from something trivial like assessing the current trend in fashion or the newest fad in technology to more important matters such as the state of the national economy.

The system is composed of 100 agents, each of whom has an opinion (assessment) on the current state of affairs, which is represented by a continuous, dynamical variable S_i (where i denotes agent number) within <0.0, 1.0>. The objective real-world situation, S_w, is also a continuous variable in the range < 0.0, 1.0>. Since we wanted to check how well the group optimizes its opinion formation, at the beginning of the simulation, we draw the agents' opinions from only a fraction of the full opinion continuum, <0.8, 1.0>. Then, we draw the world state from one of 2 ranges – far from the initial group opinion <0.0, 0.2>, and close to group opinion <0.8, 1.0> – generating two experimental cases of difficult and easy optimization tasks, respectively.

The agents are connected via directed, weighted links into a small world-network. The density of the network varied in simulation runs from 4 to 16 average links per node, and the rewiring probability from 0.1 to 0.8. A link represents a trust relation between agents – that is, the willingness of a given agent to trust and be influenced by the opinion of a particular another agent (i.e., his or her trustworthiness as perceived by that agent). Since such trust relations are rarely symmetrical in real social systems, the links in the model are directional. Their weights, w_{ij}, are randomly chosen at the beginning of the simulation from <0.0, 1.0>.

We also differentiate between trust relations (specific to the trustee, denoted by weighted, directional connections) and trustfulness, a psychological variable characterizing each agent. However, we are interested in contextually determined trustfulness that may change from situation to situation, not a stable personality trait. Therefore, the group members are characterized by a continuous dynamical variable T_i denoting their current trustfulness, initiated at 1 (fully trustful) at the beginning of the simulation.

3.2.2   Opinion Dynamics

In each simulation step, in random order, every agent adjusts his or her opinion, either by consulting with other agents with whom he or she is connected or by checking the objective state of affairs (S_w). This latter action emulates searching for low-level, factual information from sources external to the group to form an unbiased or a much less biased assessment. "Reality checking" is implicitly connected with a larger cost than is the almost negligible cost of seeking influence from one's close social circle. Such cost is due to the greater effort, longer time, or financial costs needed to obtain, gather, and to process low-level information. Although there is no implemented limit on agents' resources, each reality check is assumed to cost 1 resource unit, while accepting advice is considered costless.

The decision of whether to trust other agents' opinions or to check the state of affairs depends on the agent's trustfulness. Higher trustfulness makes the agent more likely to be influenced by his or her neighbors in the network, while lower trustfulness increases the likelihood of a reality check:

$$Action_i^t = \begin{cases} SocInfl & rand < T_i \\ RealCheck & rand \geq T_i \end{cases}$$

(3.1)

If the agent decides to rely on the opinions of others, he or she adjusts his or her assessment by summing up the weighted influence of his or her neighbors:

$$S_i^{t+1} = S_i^t + \sum_{j}^{n_j^i} \left(\left(S_j^t - S_i^t \right) \times w_{ij}^t \right)$$

(3.2)

where n_j^i denotes the number of agent's i neighbors.

If the agent decides to check the situation, he or she adjusts his or her assessment by shifting his or her opinion towards the currently verified state of affairs:

$$S_i^{t+1} = S_i^t + a \left(S_w^t - S_i^t \right)$$

(3.3)

where a is a constant denoting the agent's trust in his or her competence to gather and process information in a particular area of expertise. For the reported results $a = 1$.

Additionally, there is a small probability in each simulation step ($pRCh = 0.01$) that a random agent performs a reality check and makes an appropriate opinion adjustment apart from the regular opinion dynamics.

3.2.3   Trust Dynamics

We consider both trust relations and trustfulness to be dynamical variables that change depending on the situation and social context. The mechanics of trust and trustfulness dynamics reflect the main assumption of our approach – that coherence of opinions within a group strengthens mutual trust and decreases individuals' vigilance.

In the model, in each simulation step, agents assess the internal consistency of their interaction partners. If the opinion of a neighbor changes substantially in comparison to the previous time step, the trust link towards such an advisor drops proportionally to this change. If the neighbor is consistent in her opinion, the weight of the trust link increases slightly.

$$w_{ij}^{t+1} = \begin{cases} w_{ij}^t + \delta & \left| S_j^t - S_j^{t-1} \right| \le m \\ w_{ij}^t - \left| S_j^t - S_j^{t-1} \right| & \left| S_j^t - S_j^{t-1} \right| > m \end{cases} \tag{3.4}$$

where δ is the trust change constant, set at 0.01 and m is a sensitivity parameter that corresponds to the latitude of acceptance – that is, it describes the agent's sensitivity to the difference of opinions. This parameter thus provides a threshold on what the agents perceive as similar (indistinguishable) opinions and what constitutes a discrepancy in opinions. In real social systems, such sensitivity might be related, for example, to the importance of the matter at hand – for trivial matters, opinions or assessments even far apart could be categorized as similar, while for important decisions even slight discrepancies are noted. The hysteresis in the change of trust is implemented to mimic the catastrophic changes in attitudes observed in psychological studies (Latané and Nowak 1994). Positive attitudes grow incrementally, but the change to a negatively valenced attitude is abrupt (Jacoby et al. 2002).

The changes of agents' trustfulness are performed in each simulation step, after the opinion update. Trustfulness depends on the coherence of incoming signals an agent receives. If most of the agent's neighbors nudge him or her to change his or her opinion in a single direction (increase or decrease), he or she considers their opinions coherent. If they differ in the direction in which they push him (i.e., half of them advises an increase and half a decrease), he or she concludes they are incoherent. The agent perceives an opinion to be different from his or her own when it is larger or smaller by at least the breadth of his latitude of acceptance:

$$neighborCoh_i = \frac{\left| n^i_{j+} - n^i_{j-} \right|}{n^i_{j\pm}}$$

(3.5)

where $n_{j+}{}^I$ is the number of agent i's neighbors for which $\left(S^t_j - S^t_i \right) > m$; $n_{j-}{}^I$ is the number of agent i's neighbors for which $\left(S^t_j - S^t_i \right) < -m$; and $n_{j\pm}{}^I$ is the number of agent i's neighbors for which $\left| S^t_j - S^t_i \right| > m$.

The value of the latitude of acceptance of the agents (m) was constant during each simulation run, identical for all agents and systematically varied in different simulation runs.

If the perceived coherence drops below a set threshold, the agent's trustfulness drops substantially, proportionally to the incoherence; if it stays within the threshold, it increases gradually with each step:

$$T^{t+1}_i = \begin{cases} neighborCoh^t_i & neighborCoh^t_i < \gamma \\ T^t_i + \beta & neighborCoh^t_i > \gamma \end{cases}$$

(3.6)

where β is the trustfulness change constant, set at 0.01 and γ is coherence threshold constant, set at 0.5.

3.2.4 Results Model 1

We tested how the average opinion of the group stabilized in two conditions – when the initial average was far from the actual situation (S_w) and when it was similar. In both conditions, surprisingly, the structure of the network of trust relations (i.e., rewiring probability p and average node degree) did not impact the results. Similarly, the threshold of coherence (γ), which described what coherence of the group was sufficient to maintain trustfulness did not impact the resulting opinion dynamics. Changing the speed of changes of the dynamical variables of trustfulness and trustworthiness (link weight) – β and δ – resulted in more or less volatile opinion dynamics but did not qualitatively influence the results.

In the easy optimization condition, the results are consistent with expectations – the group quickly arrives at the average of their initial opinion and never strays from it. This result reflects the fact that even if the coherence of their relations initially fell below the threshold for some agents and they resorted to a reality check, their resultant opinion was not very far from the rest of the group and not very far from his or her previous opinion. Thus, such an agent remains trustworthy and is still a source of influence. However, the majority of agents do not perform a reality check and thus do not converge on the mean opinion in the group. Those who had their opinions influenced by a reality check are in the minority and quickly switch ro relying on social influence because the opinions of their neighbors are very coherent. In effect, only some minor changes happen in the early steps of the simulation, but the shared reality is quickly agreed upon, and from that moment, full coherence is reached. The side effect in this process is that the weight of the links – perceived trustworthiness – grows until it reaches the maximum value of 1. The group remains cohesive, and the error of group assessment is very low and depends only on how far from the mean group opinion was the randomly drawn value of the actual state of affairs.

In the second condition – difficult optimization task, where the initial state of affairs varies significantly from initial group opinions – the dynamics are far more complex. Here, the crucial control parameter becomes the opinion acceptance range, that is, the difference above which agents decide that the opinion of others is different from their own. When this parameter is high – that is, agents regard opinions even very far from their own as similar – the opinion dynamic resembles the case of an easy optimization task. A few agents might initially perform a check on what the actual situation is and change their opinion. However, they are in a small minority – as the width of the opinion acceptance range grows close to the initial range of opinions, the chance to encounter incoherence in the opinions of neighbors is minimal. Thus the rest of the group quickly stabilizes on their average opinion (Fig. 3.1), thus become coherent and in effect pull these few outsiders back into the shared group reality. In effect, the trust links strengthen and the group's opinion is stable, although it is very far from what reality looks like. We might interpret this situation as a groupthink process – the agents trust each other highly and thus reject any opinion that is far from the group average and in effect maintain a totally inaccurate world view. Wide opinion acceptance range can be interpreted as a latitude of acceptance for issues of low importance – in such cases, the differences in the opinions of one's social links are perceived as fairly irrelevant.

In the situation when opinion acceptance range is low – that is, agents perceive opinions even close to their own as dissimilar which can be the case for very important issues – the opinion dynamics are quite different. Even at the very beginning of

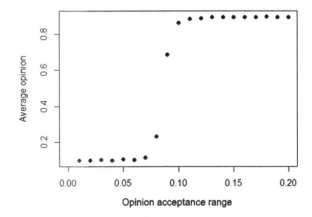

Fig. 3.1 Average opinion of the group after 800 simulation steps for different widths of the opinion acceptance range in the difficult optimization task (the actual state of affairs is drawn from <0.0, 0.2> while the initial group opinions are drawn from <0.8, 1.0>). Each point represents an average of 30 simulation runs

the simulation run, many agents find that their neighbors hold divergent opinions and thus perceive the group to be incoherent. Their trustfulness drops and they resort to a reality check – they adjust their opinions to better match the actual situation which is far from their initial one. Because of the low width of the opinion acceptance range, there is enough of such agents to make an impact on the rest of the group. They are not in a small minority and when they change their opinions they introduce additional incoherence into the group. In effect, the agents whose neighbors might have been initially very coherent now perceive far more incoherence. If among their neighbors the initial fact checkers are approximately one half of the links, they too realize that the group cannot be trusted. Thus, they check the situation and adjust their opinion based on the reality check. This process of opinion dynamics is highly non-linear – first there are a few agents changing their mind, this induces a few others to follow suit and in a few steps half of the group has opinions very different from the other half (those that checked the facts have opinions similar to the actual situation (drawn from <0.0, 0.2> and the rest have opinions close to the initial group average between <0.8, 1.0>). From there on the opinion change is very abrupt. If the opinions are split roughly into two groups, all the agents have on average half their neighbors inciting them to keep their own opinion, and the other half inciting them to change it. Thus all agents reach the maximum value for the group incoherence, which instantly – in one or two simulation steps – makes them distrust others and switch to individual information processing based on reality checks.

We can interpret this situation as a collapse of group cohesion. Each individual quickly becomes alone in his or her opinion formation and decision making, not taking any advantage of the processing capabilities of their social relations. At the same time, each of these individuals – at the cost of investing their own resources and capabilities – arrives at an opinion that is very close to the actual state of affairs. In effect, if we compute average group opinion it will also be very accurate (Fig. 3.1). As a side effect of each agent checking the facts for him or herself, the coherence of the group's opinion also increases. This gradually causes the agents to trust each other again and the strength of the connections among them grows. In the end – even though initially group cohesiveness dramatically decreased – the group arrives again at a state of full trust. This can be understood as a positive situation – high accuracy of group assessment and high trust – but this interpretation might be misleading. If any change of the real situation should happen, it might not be correctly assessed by the group, because both members' opinions and trust relations now closely resemble those in the groupthink dynamics described above. Even if a single agent from time to time change his or her opinion, he or she would not introduce enough incoherence to induce the group to reassess their judgment. In effect, this dynamics might be dysfunctional in the long run.

For intermediate values of the opinion acceptance range – precisely, for values close to the value of half of the initial group opinion range – there seems to occur a phase transition between the two types of dynamics. Depending on the precise state of initially drawn opinions and link partners, the group either follows the groupthink dynamics or the collapse of the collective dynamics.

In sum, this initial formulation of the model produces two regimes of group dynamics. For all easy optimization tasks and for difficult optimization tasks in low importance judgements, the system falls into a groupthink, which can result in very low accuracy. For difficult optimization tasks in important judgments, the group cohesiveness falls apart and each individual arrives at an opinion by herself. This can lead to an initially correct group assessment (albeit at a very high cost to group members) but can be dysfunctional in the long run.

3.3 Implementation of Model 2

The assumptions of RTSI allowed us to reproduce the two extremes in group information processing. However, there was no middle point between them and thus no optimal group dynamics. To explore the possible mechanisms of optimal group information processing, we modified the initial model in two ways.

First, we introduced changes to the objective (from the agents' point of view) state of affairs the group about which the group is forming opinions. Instead of drawing and maintaining a single value of the state of reality, we changed it into a dynamical variable that followed a sinusoidal curve with cycle length varied in simulation runs from 0.001 to 0.015 cycles per simulation step.

Second, we implemented three distinct mechanisms of link weight updating to reflect the possible ways in which individuals can change their assessment of the trustworthiness of their social relations. In different simulation runs, we varied the trustworthiness update algorithms in order to compare their effectiveness and impact on group information processing.

These three rules for changing the strength of trust relationships depended on the internal consistency of the trustee, continuous similarity to self, and similarity to self only at moments of high self-assurance. For each, we have systematically varied the speed with which the reality changes to check how well the social group adapts and the width of the opinion acceptance range.

The first mechanism – consistency – was identical to the implementation of trustworthiness changes in Model 1, as described in Eq. 3.4. Thus, agents' trust towards their neighbors depended on how consistent they were. This mechanism can be related to the natural tendency to trust those who stand by their opinions and mistrust people who cannot make up their mind. If a person changes their opinion from day to day, it is unlikely that they will be trusted by others and asked for advice. In comparison, stable opinions are more likely to induce trust from others.

In the second mechanism, trust towards a particular neighbor depends on the similarity of opinions between the agent and this neighbor. This mechanism is based on the observation that we tend to trust people who are similar to us and distrust those who represent vastly different opinions. Therefore each agent in each step and for each of the neighbors assesses how their opinions vary from his or her own.

$$w_{ij}^{t+1} = \begin{cases} w_{ij}^t + \delta & \left|S_j^t - S_i^t\right| \le m \\ w_{ij}^t - \left|S_j^t - S_i^t\right| & \left|S_j^t - S_i^t\right| > m \end{cases} \tag{3.7}$$

The final version is similar to the second one in that trust depends on the similarity of opinions but this time the updates of trust links are performed at specific time points. Trust relations change when agents have the opportunity to compare the assessment of their neighbors with the actual state of affairs – that is, after performing a reality check. This mechanism resembles situations where after gaining external, objective information, individuals are confident of their opinions and reassess the competence of their social contacts in the given area of expertise. For example, if a recently bought dishwasher – of a model highly valued by a close friend for its durability – malfunctions, an individual might conclude that the friend is not an expert on houseware, so that when buying another appliance, he or she might disregard advice from that friend.

In the model, after performing a reality check and adjusting their opinion accordingly, agents increase or decrease the weights of their trust links depending on whether their neighbors are close enough to their reality-informed assessment:

$$w_{ij}^{t+1} = \begin{cases} w_{ij}^t + \delta^I & \left|S_j^t - S_i^t\right| \le m \\ w_{ij}^t - \delta^I & \left|S_j^t - S_i^t\right| > m \end{cases} \tag{3.8}$$

where δ^I is the trust change constant, set at 0.2 and m is a sensitivity parameter, which describes agent's sensitivity to the difference of opinions.

3.3.1 Results Model 2

To assess how well the modelled social system optimizes group opinion, we were interested in the balance between the accuracy of the group's assessment and the cost of arriving at it. A group that is accurate at a high cost is not well optimized, as is a group who bears little costs but is not accurate – just as were the groups in Model 1. We consider the dynamics to be optimized when the accuracy is relatively high while the costs are kept low. We measure the accuracy as the difference between an agent's opinion and the state of affairs, averaged over all group members. The cost is measured as the fraction of group members that performed a reality check in a given simulation step. Unlike Model 1, the agents do not stabilize on a single opinion value as the world situation is changing. Thus, the measures of optimality were averaged over 1000 simulation steps.

We have analyzed how such defined optimization depends on: structure of the network of trust links (density, rewiring probability), properties of the individuals (their opinion acceptance range that determines sensitivity to differing opinions), and properties of the world (the speed of changes of the objective state of affairs). Each of these variables was tested for 3 methods of trust links dynamics – (a) opinion change (i.e., the consistency of the trustee), (b) opinion similarity, (c) opinion validity (i.e., the difference between trustee's opinion and reality). In all cases, we drew the initial opinions similarly as in Model 1 (from a segment of possible opinions <0.8, 1.0> to mimic a fairly cohesive group and set initial link weights to random to differentiate group structure.

3.3.1.1   Network Structure

When the trust links are modified depending on the objective assessment of situation, the group's optimality does not depend either on network density or rewiring probability. For the other two cases, the group error grows (and the cost drops) for networks with more than 8 links per node on average and for rewiring probability larger than 0.2. Similarly, below those values the group dynamics also seems suboptimal: the lower the rewiring probability the higher the cost and the lower the error. Lowering average node degree for the networks that update trust link strength based on neighbors' consistency leads to the same outcome – low error but high cost. For those networks where links are updated continuously based on opinion similarity, the situation is slightly better – the error stays fairly stable when lowering the average degree below 8 and the cost drops.

In sum, network structure had a surprisingly modest influence on the group's optimality. The balance point (where accuracy was best matched with cost) was determined at density of 8 links per node and rewiring probability of 0.2. For further analyses, these network settings were used as default (Fig. 3.2).

3.3.1.2   Speed of Changes and Opinion Window

The speed of changes and agents' opinion window influence group optimization in different ways. As Fig. 3.3 shows, the speed of changes has very little impact on group opinion accuracy, no matter which way the trust links are updated. However, the faster the changes the higher the cost, measured by the fraction of group members that need to investigate the facts. What this means is that all of the investigated structures were able to maintain accuracy no matter how fast the situation changed but at higher speeds such accuracy was more costly.

Opinion acceptance range (i.e., the sensitivity to differing opinions) influenced both the group error and the cost. For all trust links update modes, the wider the opinion window, the higher the error and the lower the cost – similarly to the switch between the regimes of collective collapse and groupthink in Model 1. However, this relation was not linear. For very narrow opinion windows (i.e., situations of high importance, when even slightly higher or lower opinion was considered different from the agent's own) the cost was very high (almost 40% of members needed to verify the situation in each step) and the error was very low (close to 0). For very

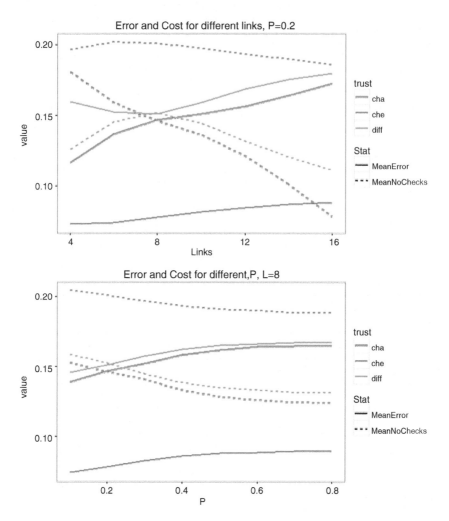

Fig. 3.2 Impact of network structure on cost and accuracy of group opinion. Each point represents an average of 30 simulation runs, averaged over 1000 simulation steps and all opinion acceptance ranges (coherence sensitivity) and all speeds of changes (cycles) but for a single value of average node degree (a, L = 8) and rewiring probability (b, p = 0.2). Colors represent different link update mechanisms: 'cha', red – based on trustee consistency, 'che', blue – based on trustee's opinion similarity to the reality, 'diff', green – based on trustee's similarity to the agent. Solid line represents group opinion error and dashed lines represent the cost.

broad opinion windows the situation was reversed – the error was high and cost very low. For each trust link update mode, there was an intermediate width of opinion window for which both measures were at moderate, balanced levels. However, the modes differed in where such balance was achieved: for trust dynamics based on opinion similarity and trustee consistency this happened for opinion windows between 0.08 and 0.09; and for trust dynamics based on reality assessment between

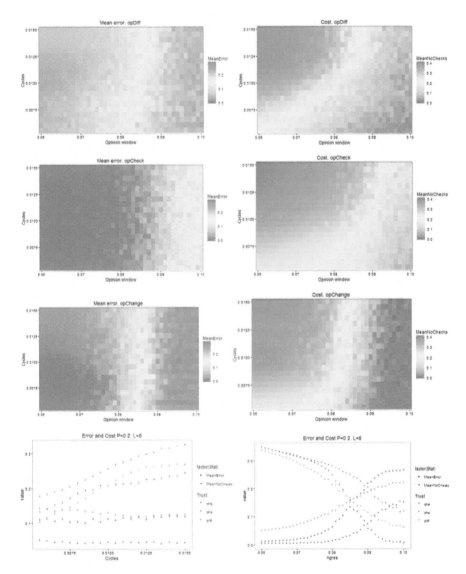

Fig. 3.3 Mean group error and cost as a function of the agents' opinion acceptance range (for simplicity, labeled on the images as 'opinion window', X axis) and speed of changes of the objective state of affairs (in cycles per simulation step, Y axis). The bottom row compares these results for different trust update mechanisms

0.09 and 0.1. Moreover, the modes differed also in the shape of this relation: when trust was updated based on opinion similarity, the dependence between error, cost and opinion window was almost linear. When it was updated based on trustee consistency, the relation was the least linear, with an abrupt switch between low error with high cost to high error with low cost – just as could be observed in the phase transition in Model 1 (Fig. 3.1). Trust updated by reality assessments led to a relation that can be described as a middle point between these extremes.

To see what type of group dynamics is present in each of these three regimes of opinion acceptance range, we analyzed what end results of mean group error constitute the averages presented in Fig. 3.3. Is it truly a phase transition – that is, are only 2 regimes are present or is there a middle point? We have computed the entropy and histograms of mean error from the 30 simulation runs that were averaged to arrive at the data presented in Fig. 3.3. As can be seen in Fig. 3.4, the mean error in trustee consistency mode of trust link updates is very predictable below .08 opinion window width and above .09. For trust links based on reality assessment, the predictability is very high below .09, whereas for trust links updated based on opinion similarity, predictability is generally low. This means that in this condition from simulation run to simulation run, the mean error averaged over 1000 simulation steps could be anywhere from very low to very high. In the other 2 conditions, this unpredictable behavior happened only within certain limits.

Figure 3.5 presents detailed information on what constitutes this unpredictability in the most interesting combinations of cycles and opinion window parameters. As can be seen from the probability density graphs, for trust links based on opinion similarity and reality assessment, the unpredictability comes from almost equal probability of any result of mean error for the given simulation run. For the condition of trustee consistency, this effect comes from a bimodal distribution – either the error was very low or very high. Interestingly, this unpredictability regime for this condition resembles a phase transition – the dynamics of the system does not gradually switch from low error to high error but instead is bi-stable. This result is very similar to the results of Model 1 and represents a situation where there is no middle point between the two suboptimal group dynamics.

It is worth noting that in order for the mean error averaged over 1000 simulation steps to be low, it has to decrease very early and remain low even when the objective state of affairs changes. Similarly, for the mean error to be high, it has to be high from the beginning to the end of the simulation. It is clear, then, that for the condition of trustee consistency the system either does not follow the real world situation at all (i.e., the error is high from start to finish) or immediately at the beginning switches to following the reality and keeps on doing that for the whole simulation – through investment of high costs, that is, many agents need to constantly check the facts. Above opinion window of 0.09, only the first dynamics is possible, below 0.08 only the second is possible, and in between these boundaries both are almost equally probable. In effect, the nonlinearity of the averages presented in Fig. 3.1 (Model 1) and Fig. 3.2 (Model 2, for the trustee consistency mode) comes from the fact that in this system only two types of dynamics are present.

For the other two conditions, the situation is less clear, as the averages in Fig. 3.3 seem to be computed from almost flat distributions in the unpredictable regime. To see what dynamics constitutes these effects, we analyzed sample simulation runs.

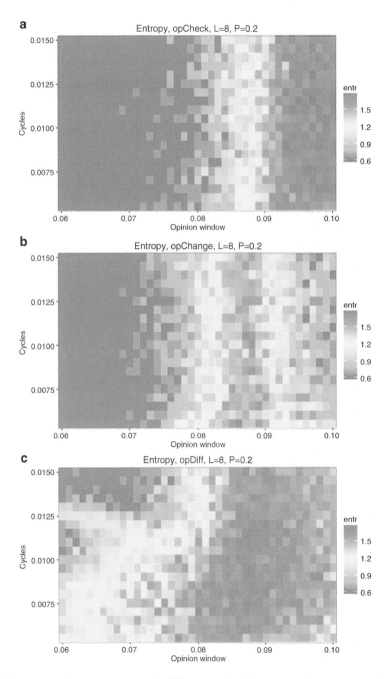

Fig. 3.4 Entropy of mean group error for different modes of trust link updates: (**a**) trustee's similarity to the objective state of affairs, (**b**) trustee's consistency, and (**c**) trustee's similarity to the agent

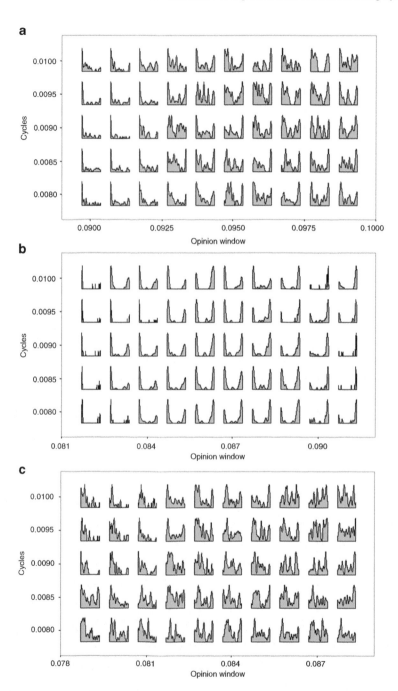

Fig. 3.5 Histograms of mean group error for different modes of trust link updates: (**a**) trustee's similarity to the objective state of affairs, (**b**) trustee's consistency, and (**c**) trustee's similarity to the agent

3.3.1.3 Optimal Dynamics

For moderate values of sensitivity to difference, the most interesting dynamics can be seen. Initially the agents perform reality checks and adjust their opinions accordingly, but once most of the group has verified the state of affairs, trustfulness and trust grow and agents start to rely more on their neighbors' opinions. When the real situation changes, a small set of agents checks the reality, with the rest following their assessment rather than performing a check themselves. The group error fluctuates slightly but on average is moderately low. In this regime of dynamics, reality checks are balanced with social influence (around 30 agents checking the situation in each simulation step).

To understand the optimal group dynamics that preserves social relations and social influence but at the same time keeps the group decision process accurate, we can analyze a sample simulation run in either of the two types of trust link changes that exhibit the middle point between groupthink and collapse of collective (trustee's opinion similarity to facts, and trustee's similarity to agent). An example can be seen in Fig. 3.6.

The first observation is that both the mean error and the group cost (number of agents performing reality checks) fluctuate, following roughly (with some delay) the reality changes (green line in Fig. 3.6a). However, the two crucial parameters influence these dynamics in different ways. Sensitivity to difference determines the minima and maxima of cost fluctuations, with higher importance shifting the whole trajectory to higher values. The speed of changes, on the other hand, influences the amplitude of cost fluctuations, with higher speeds causing the amplitude to decrease. This interesting pattern is due to the fact that agents need some time to "realize" that the situation has changed (i.e., the derivative of the sinusoid has changed sign). With low speed of changes, this causes a certain delay of agents' reality checks: at first only few check the situation, this causes incoherence in their neighbors' perception and reality checks on their part as well. The severity of this cascade depends on the sensitivity parameter. However, if the speed of changes is very high, the initial cascade is shortly followed by another, as agents keep on reality checking and adjusting their opinions. At some point, this process causes the cost fluctuations to disappear and instead a stable fraction of agents checks the situation at each step.

The mean group error (which can be interpreted as the accuracy of group decisions) is strongly dependent on the number of reality checks at each step. In the case of optimal group dynamics, it fluctuates at the beginning (in the first few reality change cycles) to finally settle at very low levels and remain fairly stable. This reflects the dynamics of reality checking– when the group first realizes that the situation has changed (i.e., when a sufficient number of agents checks the reality and incoherence of opinions starts to be substantial), very many agents perform a reality check and this causes the error to drop almost to zero. Subsequently, since the coherence is restored, for a long period (half a cycle) almost no one checks the situation and the error grows. However, with each cycle, the number of reality checks gets better adjusted to the speed of changes: a larger number of agents check the world situation at each direction change (peak of the cycle) to subsequently keep the

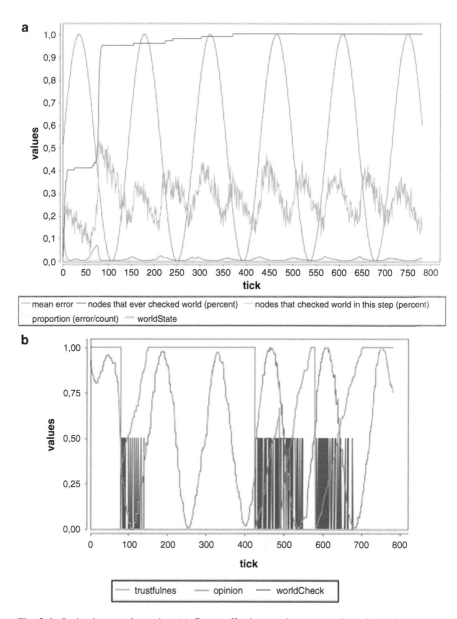

Fig. 3.6 Optimal group dynamics. (**a**) Group effectiveness in an exemplary simulation run: the mean error decreases quickly (red) even as the state of the issue changes (gray) through a moderate number of agents' information seeking behaviors (green for each time step, blue – cumulative). (**b**) Exemplary node in the same simulation run as (**a**): one node's opinion dynamics (gray) follows the issue changes (gray line in **a**) in some cycles by relying on opinion of others when the trust (red) is high and in some on own resources when trust is low (blue lines denote reality checks)

error low through social influence and only a minimal number keep checking for the remainder of the cycle. This small number of reality checks is crucial so that at the next cycle peak, these agents can "warn" the group of the change through introduction of incoherence of opinions. This process of fitting the group dynamics to the speed of changes causes the costs to fluctuate slightly with change cycles but keeps the error at a low value throughout the whole cycle.

3.3.1.4   Group Roles

Analysis of a sample simulation run shows some emergent properties in the group dynamics: the agents differ in their temporal patterns of reality checks. Even though there is no heterogeneity in the most crucial variable describing the agents – sensitivity to opinion difference – variations in checking behavior still show. There are three patterns of reality checks: no checks or only a few throughout the simulation run, checks appearing in almost every step and bursting pattern in which checks appear at each step for some time and then disappear for another period (Fig. 3.6b). These patterns can be easily related to any psychological variable that determines suspiciousness but in this simulation they are not a result of agent properties. Rather, they are an effect of agent's placement in the social network, that is., link strength, number, neighbors (which is randomized, so in each simulation these patterns appear for different agents).

The "no-checkers" are the free riders in the social system – these are the people that do not invest in checking but through proper positioning in the network still have an accurate opinion. For example they could be linked to mostly agents that check often and therefore are coherent and accurate. We could view this as a negative trait (i.e., abuse of others) but it can be also said that these are the people in the network with the strongest social capital. They could be, for example, top managers or chief organizers that gather information from dedicated social contacts to arrive at a binding decision guiding the whole organization.

The "bursters" are the biggest group (Fig. 3.6b). They differ in the length and frequency of bursts and therefore represent the majority of social group members who differ in where they put the balance between investment of own resources and social influence. This again depends not on agent's characteristic but on the quality of their social ego network (their direct connection to others). Their links could be many and varied and therefore prone to display some incoherence from time to time.

Finally, the "checkers" are the most interesting group. The easiest interpretation might be that they are those who are naïve enough to be doing all the hard job for the rest of the group. However, in fact it is again their position that matters – they may have diverse links and just a few – to others from the same group and to those that rely mostly on social influence. For example, with only 2 neighbors, it only takes one to differ to introduce full incoherence. With a small number of close links, if there is any incoherence, for them it would seem huge – they instantly notice it.

This is what keeps the group on track. They are the devil's advocates who question the dominant opinion. Because they check all the time, they are the first to spot any difference between what comes through social links and the real state of affairs. Without this initial recognition, there would not be enough incoherence to push more members into checking and the group would go astray.

What is crucial from these descriptions is that the roles in the group come only from the network structure – the number of neighbors most of all. Even if they are not much trusted (i.e., the strength of the connection is low), they count into the agent's closet environment and serve as a gauge of the distribution of opinions in the group. Agents with many neighbors will need many of them to change their opinion in order for incoherence to appear. The hubs in the network might never reach that point. On the other hand, those who only have few connections will be more sensitive to any change in opinions. In the middle are those who have moderate number of links in which incoherence happens from time to time only.

3.4   Discussion

The results of the models reveal three distinct types of dynamics depending on the importance of the assessed issue to the agents and the speed at which the objective state of affairs changes. The first dynamics type is a total collapse of the social structure, in which the group reduces their mean error of assessment but at a great cost. Each agent works for him or herself and trust drops dramatically. The second dynamics type represents groupthink – the group converges on a unified opinion that is detached from the real state of the issue. The third – optimal – dynamics type is such that the group collectively decreases its mean error of judgment at a relatively low cost (only selected individuals search for information).

The intuition behind the models' mechanics and implementation was that in social systems, individuals try to optimize their decision making by balancing advice seeking (i.e., social influence) and information processing. When trying to arrive at an opinion or making a choice, one may either rely on one's social contacts – and their knowledge – or invest time and effort to form an informed assessment oneself. The first option allows for time-saving but carries an inherent risk that the social contacts' opinions may be biased or inaccurate. The latter approach promises more accuracy but at a greater personal cost. To optimize costs and effectiveness of this process, individuals have to rely on some social cues to guide their choices whether to trust their social links or to seek low level facts.

Coherence of opinions seems a natural choice for such a social cue. Intuitively, if a large number of experts or otherwise (i.e., morally) trustworthy sources agree on a matter, there is high probability that the reality does not stray far from their assessment. This premise can be derived from the mechanism of the so-called wisdom of the crowds: the average opinion of many, varied and independent judges is most often closer to the truth than any single judgment alone. From this it seems to naturally follow that if a large number of opinions are coherent (they are all close to the average), they should be accurate.

However, there is one catch in this logic. For the wisdom of the crowds to work, the opinions of the judges should be independent of each other; more specifically, the error terms must not be correlated (Page 2008). In reality, many people seem to assume their opinion sources to be independent unless there are clear signs to the contrary. For example, if a political or economic issue is discussed by a panel of experts, we conclude that they voice independent judgments. However, if we hear two members of one political party arrive at the same opinion – even if it happens in separation – we usually conclude that they state the dominant stance of their party's program.

The situation is much more complex in the case of opinions spreading through social media. With the sheer volume of incoming information constantly growing, it becomes increasingly difficult to verify opinion sources against any factual (or at least less biased) data. More importantly, it is almost impossible to trace whether a particular statement arriving from many social links simultaneously originated from a single opinion leader or is an opinion shared by many. News sites usually state their sources, but they do not constitute any longer the most prominent information source – opinions are increasingly based on social networks information flows.

In effect, we often fall into the trap of assuming the coherence of opinions to be a reliable cue of their accuracy, while in reality it mainly stems from the dependence of the sources – that is, the same information spread and repeated by many social connections. We hypothesize that this mistakenly applied heuristic is responsible for the occurrence of the groupthink phenomenon, visible in social media as echo chambers. Groupthink is usually described in very specific conditions (small groups, important decisions, clear leadership), diagnosed in case studies only and rarely researched in experimental settings. It occurs when the pressure towards uniformity is so large that people holding divergent views tend to refrain from sharing them with the group in fear of being ostracized. The group as a whole becomes sheltered from controversial information, which may lead to a distorted perception of facts (Janis 1972).

We observe however that the main consequence of groupthink – the detachment of group opinions and decisions from reality – may occur in many settings through a positive feedback loop of increasing coherence and trust. This can occur in social systems mediated through ICT in the form of echo chambers when the individuals use the homophily criterion in choosing their friends and joining social groups (McPherson et al. 2001).

When surrounded by similar peers, sharing the same attitudes, values and interests, people only have access to information that confirms their own point of view. Thus, the social network starts to resemble an echo chamber in which some ideas are being repeated and amplified, and some (usually discrepant) are censored or simply underrepresented. This effect is even stronger in social groups that interact on social media platforms. There, the natural tendency to mix with similar others is strengthened by algorithms which provide the content that is discussed and shared and which aim at increasing the content's ratings or "likes" (Bakshy et al. 2015), in effect creating a "filter bubble" for the group (Pariser 2011).

The result of these processes is that individuals think that their own personal outlook is more widely accepted than it really is. On the other end of the spectrum, there are people who feel that their opinions are not represented within a social system. According to the spiral of silence theory (Noelle-Neumann 1974), they tend to refrain from voicing their opinions due to the fear of being isolated or excluded from the group. As a result, the controversial idea becomes even less represented in public opinion. Both effects are results of positive and negative feedback loops that govern the dynamics of information flow in social systems.

One of the traps of group cognitive processing is the press towards unanimity or social conformity. When group members are confronted with opinions contrary to their own, they tend to shift their judgments towards the view of the majority (Asch 1955). The need to conform to the prevalent group opinion may evoke self-censorship in expressing overt dissent (Janis 1972). Group members may feel that they will be punished for expressing opinions that are not in line with the group consensus. This may have fatal consequences, as in the Bay of Pigs disaster (Janis 1972), or more recently, the Tenerife airport disaster (Haerkens et al. 2012). One of the remedies against negative effects of group conformity is a deliberate introduction of a devil's advocate – a group member whose task is to adopt a position that is divergent from the prevailing group norm. The role of a devil's advocate is to raise objections to preferred alternatives, challenge assumptions adopted by the group, and propose different points of view (e.g., de Bono 1985; Schwenk 1984). Many studies show that introducing this role playing technique in group decision making significantly improved the quality of group decisions (Cosier 1978; de Villiers et al. 2016; Herbert and Estes 1977; Schweiger et al. 1989; Schwenk 1984, 1990). By making critics accepted and valuable, the group may protect itself from relying solely on commonly agreed and accepted knowledge (Sunstein and Hastie 2014). Our models show that even a small number of devil's advocates can help the group remain accurate while minimizing the costs.

Epilogue

Our aim in preparing this volume was, in one sense, quite limited. We set out to identify a few variables that interact to shape how and why individuals seek out other people for influence in making decisions and forming judgments about issues and events. But in another sense, our aim was very ambitious. We attempted to establish, in minimalist terms, the common denominator underlying the fundamental processes that combine to generate the whole spectrum of ways in which the individual is related to others in a social system. This lofty goal, though, should be viewed as an attempt to create a scaffold that sets the agenda for future research to pursue. In this sense, the theory and research we have presented should be viewed as heuristic rather than comprehensive.

Like any heuristic model, what we propose in this book is open to tweaking and possible subject to revision, as new empirical research critically examines and builds upon the ideas we have presented. The scientific literatures on social influence and group dynamics are extensive, having been generated over several decades, so it is naïve to expect that a few simple rules cannot capture the complexities and intricacies of all the empirical findings.

In an important respect, our theory pays homage to the classic insight of Allport (1954), who argued that the basic agenda of social psychology is to understand how the thoughts and actions of an individual are influenced by the real of imagined action of other people. Allport's framing of social psychology as the study of social influence is widely acknowledged by everyone, but widely under-appreciated by the very scholars who appreciated his insight. Indeed, when social psychologists discuss social influence, they confine their efforts to investigating how people manipulate one another and attempt to overcome the resistance of others in doing so. Rather that looking at how social influence underlies social phenomena of all kind, most psychologists have defined influence as essentially synonymous with manipulation, involving strategies that are overt (e.g., obedience) or subtle (e.g., compliance strategies. Asymmetrical influence in service of achieving personal goals in a zero-sum sense is hardly what Allport had in mind.

The RTSI perspective elevates social influence by suggesting that it is the fundamental underpinning of the individual in his or her attempt to navigate multiple forces in social life. Influence, broadly considered, goes to the very heart of complexity science. Thus, any complex system involves influence among the elements comprising the system. In a social system, the elements are individuals that influence each other through verbal and non-verbal means. So what looks like asymmetric manipulation or overcoming resistance when only two elements are singled out," misses the forest for the trees." When one broadens the scope to view the system more globally, one recognizes that asymmetric influence is a special case of the widespread influence that percolates throughout the system. Indeed, were it not for the system-wide nature of influence, one could argue that the system would not exist in the first place.

We have argued that social groups can be looked upon as highly efficient distributed information processing systems. This is not to deny the vast and diverse lines of theory and research exposing the problematic nature of influence in groups of all kinds. Research has shown, for example, that judgments produced by groups of interacting individuals are less accurate than the average of judgments of independent individuals (Lorenz, Rauhut, Schweitzer, & Helbing 2011)). There is convincing evidence, too, that when groups of individuals with impressive expertise discuss important matters, they may converge on very poor decisions (Janis 1972). Simply forming a group and asking them to arrive at a joint decision does not guarantee that it will function as an efficient and effective information processing system.

Recognizing the potential for problematic functioning in groups, we propose that socially distributed information processing needs balanced mechanisms of regulation to function properly. In this view, trust is not just an asset or a form of social capital that leads to efficient and smooth functioning of social systems, but rather represents a crucial control parameter in the regulation of distributed social information processing. An insufficient degree of trust hinders or terminates the delegation of information processing to others, stalls the flow of information in social systems, and as a consequence, impairs the performance of individuals and groups. Too much trust, in turn, removes the negative feedback loops from the system, removes checks on information processing, and makes the system a victim of runaway dynamics of trust that promotes full reliance on unchecked and potentially false information.

Optimal individual and group functioning thus require some level of trust, but also some degree of distrust or doubt, such that there is reliance on others along with sensitivity to, and checking for incoherence that can signal malfunctioning of the system. The processing resources of individuals and groups, moreover, should be devoted to processing information concerning matters with the greatest importance. These control parameters are not static, but rather should be dynamically adjusted to accommodate changes in the system's performance and in the context. The challenge for future research is to find the optimal configuration of these key parameters and how they change across time and across conditions.

Pending the outcomes of such research, we may be in a position not only to enhance the functioning of human groups and avoid the pitfalls that have been identified in research, but also to design artificial agents that will function in techno-social systems in an efficient and effective manner. We hope the ideas we have presented in this volume help chart these research agendas. Such efforts will not only contribute to science but will also improve the cognitive and interactive abilities of human groups and societies.

References

Adorno, T.W., Frenkel-Brunswik, E., Levinson, D.J., Sanford, R.N., Aron, B.R., Levinson, M.H., Morrow, W.R.: The authoritarian personality. Norton, New York (1950)

Albert, R., Barabási, A.L.: Statistical mechanics of complex networks. Rev. Mod. Phys. **74**(1), 47 (2002)

Allport, G.W.: Handbook of social psychology. Cambridge University Press, Cambridge, MA (1954)

Apsler, R., Sears, D.O.: Warning, personal involvement, and attitude change. J. Pers. Soc. Psychol. **9**(2p1), 162 (1968)

Arrow, K.J.: The limits of organization. Norton, New York (1974)

Arrow, H., McGrath, J.E., Berdahl, J.L.: Small groups as complex systems: formation, coordination, development, and adaptation. Sage, London (2000)

Asch, S.E.: Social psychology. Prentice-Hall, New York (1952)

Asch, S.E.: Opinions and social pressure. Sci. Am. **193**(5), 31–35 (1955)

Asch, S.E.: Studies of independence and conformity: I. A minority of one against a unanimous majority. Psychol. Monogr. Gen. Appl. **70**(9), 1 (1956)

Back, K.W.: Influence through social communication. J. Abnorm. Soc. Psychol. **46**(1), 9 (1951)

Bakshy, E., Messing, S., Adamic, L.A.: Exposure to ideologically diverse news and opinion on Facebook. Science. **348**(6239), 1130–1132 (2015). https://doi.org/10.1126/science.aaa1160

Barabási, A.L., Albert, R.: Emergence of scaling in random networks. Science. **286**(5439), 509–512 (1999)

Barber, B.: The logic and limits of trust. Rutgers University Press, New Brunswick (1983)

Baron, R.S., Vandello, J.A., Brunsman, B.: The forgotten variable in conformity research: impact of task importance on social influence. J. Pers. Soc. Psychol. **71**(5), 915–927 (1996)

Bar-Tal, D.: Shared beliefs in a society: social psychological analysis. Sage, London (2000)

Baumeister, R.F., Leary, M.R.: The need to belong: desire for interpersonal attachments as a fundamental human motivation. Psychol. Bull. **117**(3), 497 (1995)

Bentley, R.A., Ormerod, P., Batty, M.: Evolving social influence in large populations. Behav. Ecol. Sociobiol. **65**(3), 537–546 (2011)

Berg, J., Dickhaut, J., McCabe, K.: Trust, reciprocity, and social history. Games Econom. Behav. **10**(1), 122–142 (1995)

Bonabeau, E.: Agent-based modeling: methods and techniques for simulating human systems. Proc. Natl. Acad. Sci. **99**(Suppl 3), 7280–7287 (2002)

Bourdieu, P.: The forms of capital. In: Halsey, A.H., Lauder, H., Brown, P., et al. (eds.) Education: culture, economy, society, pp. 46–58. Oxford University Press, Oxford (1997)

Brady, J.V.: Ulcers in "executive" monkeys. Sci. Am. **199**(4), 95–103 (1958)

© The Author(s), under exclusive licence to Springer Nature Switzerland AG 2019
A. Nowak et al., *Target in Control*, SpringerBriefs in Complexity,
https://doi.org/10.1007/978-3-030-30622-9

Brehm, J.W.: A theory of psychological reactance. Academic, New York (1966)

Bui-Wrzosińska, L., Biesaga, M., Nowak, A.: Self-regulation of social influence: distribution of attention and level of information seeking by target of influence (unpublished manuscript) (2016)

Cacioppo, J.T., Petty, R.E.: The need for cognition. J. Pers. Soc. Psychol. **42**(1), 116 (1982)

Chiles, T.H., McMackin, J.F.: Integrating variable risk preferences, trust, and transaction cost economics. Acad. Manag. Rev. **21**(1), 73–99 (1996)

Cialdini, R.B.: Influence: how and why people agree to things. Quill, New York (1984)

Cialdini, R.B.: Influence: the psychology of persuasion. Morrow, New York (1993)

Coleman, J.: Social capital in the creation of human capital. Am. J. Sociol. **94**, 95–120 (1988)

Coleman, J.S.: Foundations of social theory. Belknap Press, Cambridge, MA (2000)

Colleoni, E., Rozza, A., Arvidsson, A.: Echo chamber or public sphere? Predicting political orientation and measuring political homophily in Twitter using big data. J. Commun. **64**(2), 317–332 (2014)

Cook, K.S., Hardin, R., Levi, M.: Cooperation without trust? Russell Sage Foundation Publ, New York (2005)

Cosier, R.A.: The effects of three potential aids for making strategic decisions on prediction accuracy. Organ. Behav. Hum. Perform. **22**(2), 295–306 (1978)

Craik, F.I., Lockhart, R.S.: Levels of processing: a framework for memory research. J. Verbal Learn. Verbal Behav. **11**(6), 671–684 (1972)

Cuddy, A.J., Fiske, S.T., Glick, P.: Warmth and competence as universal dimensions of social perception: the stereotype content model and the BIAS map. Adv. Exp. Soc. Psychol. **40**, 61–149 (2008)

Currall, S.C., Epstein, M.J.: The fragility of organizational trust: lessons from the rise and fall of Enron. Organ. Dyn. **32**(2), 193–206 (2003)

Currall, S.C., Inkpen, A.C.: On the complexity of organizational trust: a multi-level co-evolutionary perspective and guidelines for future research. In: Bachmann, R., Zaheer, A. (eds.) Handbook of trust research, pp. 235–246. Edward Elgar Publishing, Cheltenham (2006)

Davidsson, P.: Agent based social simulation: a computer science view. J. Artif. Soc. Soc. Simul. **5**(1), (2002)

de Bono, E.: Thinking hats. Little, Brown and Company, London (1985)

de Villiers, R., Woodside, A.G., Marshall, R.: Making tough decisions competently: assessing the value of product portfolio planning methods, devil's advocacy, group discussion, weighting priorities, and evidenced-based information. J. Bus. Res. **69**(8), 2849–2862 (2016)

DeBruine, L.M.: Facial resemblance enhances trust. Proc. R. Soc. Lond. B Biol. Sci. **269**(1498), 1307–1312 (2002)

Deluga, R.J.: The relation between trust in the supervisor and subordinate organizational citizenship behavior. Mil. Psychol. **7**(1), 1–16 (1995)

Deutsch, M.: Trust and suspicion. J. Confl. Resolut. **2**(4), 265–279 (1958)

Deutsch, M.: Cooperation and trust: some theoretical notes. Nebraska symposium on motivation, pp. 275–320. Nebraska University Press, Lincoln (1962)

Deutsch, M., Gerard, H.B.: A study of normative and informational social influences upon individual judgment. J. Abnorm. Soc. Psychol. **51**(3), 629 (1955)

Devaney, R.: An introduction to chaotic dynamical systems. Westview press, New York (2008)

Dietz, G., Den Hartog, D.N.: Measuring trust inside organisations. Pers. Rev. **35**(5), 557–588 (2006)

Dimoka, A.: What does the brain tell us about trust and distrust? Evidence from a functional neuroimaging study. MIS Q. **34**(2), 373–396 (2010)

Dodgson, M.: Learning, trust, and technological collaboration. Hum. Relat. **46**(1), 77–95 (1993)

Eagly, A.H., Telaak, K.: Width of the latitude of acceptance as a determinant of attitude change. J. Pers. Soc. Psychol. **23**(3), 388 (1972)

Earle, T.C.: Trust in risk management: a model-based review of empirical research. Risk Anal. **30**(4), 541–574 (2010)

Ekeh, P.P.: Social exchange theory: the two traditions. Haward University Press, Cambridge, MA (1974)

Festinger, L.: Informal social communication. Psychol. Rev. **57**(5), 271–282 (1950)

Festinger, L.: A theory of social comparison processes. Hum. Relat. **7**(2), 117–140 (1954)

Festinger, L.: A Theory of cognitive dissonance, vol. 2. Stanford University Press, Stanford (1962)

Fine, B.J.: Conclusion-drawing, communicator credibility, and anxiety as factors in opinion change. J. Abnorm. Soc. Psychol. **54**, 369–374 (1957)

Fiske, S.T., Taylor, S.E.: Social cognition. Mcgraw-Hill Book Company, New York (1991)

Fiske, S.T., Cuddy, A.J., Glick, P.: Universal dimensions of social cognition: warmth and competence. Trends Cogn. Sci. **11**(2), 77–83 (2007)

French, J.R., Raven, B., Cartwright, D.: The bases of social power. Classics Organ. Theory. **7**, 311–320 (1959)

Fukuyama, F.: Trust: the social virtues and the creation of prosperity. Free Press Paperbacks, New York (1995)

Fukuyama, F.: Social capital and development: the coming agenda. SAIS Rev. **22**(1), 23–37 (2002)

Gambetta, D.: Trust: making and breaking cooperative relations. Basil Blackwell, New York (1988)

Giardini, F., Quattrociocchi, W., Conte, R.: Understanding opinions. A cognitive and formal account. Advances in Complex Systems (2011)

Gilbert, N.: Agent-based social simulation: dealing with complexity. Complex Syst. Netw. Excell. **9**(25), 1–14 (2004)

Glöckner, A., Engel, C.: Can we trust intuitive jurors? Standards of proof and the probative value of evidence in coherence-based reasoning. J. Empir. Leg. Stud. **10**(2), 230–252 (2013)

Gorn, G.J.: The effects of personal involvement, communication discrepancy, and source prestige on reactions to communications on separatism. Can. J. Behav. Sci. **7**, 369–386 (1975)

Gouldner, A.W.: The norm of reciprocity: a preliminary statement. Am. Sociol. Rev. 161–178 (1960)

Grzelak, J.Ł., Nowak, A.: Wpływ społeczny. In: Strelau, J. (ed.) Psychologia Podręcznik Akademicki, vol. 3, pp. 187–205 (2000)

Guastello, S.J., Koopmans, M., Pincus, D. (eds.): Chaos and complexity in psychology: the theory of nonlinear dynamical systems. Cambridge University Press, Cambridge (2008)

Haerkens, M.H., H, D., Van der Hoeven, H.: Crew resource management in the ICU: the need for culture change. Ann. Intensive Care. **2**(1), 39 (2012)

Hardin, C.D., Higgins, E.T.: Shared reality: how social verification makes the subjective objective. Handbook of motivation and cognition, Vol. 3: the interpersonal context, pp. 28–84. The Guilford Press, New York (1996)

Harkins, S.G., Petty, R.E.: The effects of source magnification on cognitive effort and attitudes: an information processing view. J. Pers. Soc. Psychol. **40**, 401–413 (1981)

Harkins, S.G., Petty, R.E.: Social context effects in persuasion: the effects of multiple sources and multiple targets. In: Paulus, P. (ed.) Advances in group psychology, pp. 149–175. Springer, New York (1983)

Harkins, S.G., Petty, R.E.: Information utility and the multiple source effect. J. Pers. Soc. Psychol. **52**(2), 260 (1987)

Hayashi, Y., Kryssanov, V.: An empirical investigation of similarity-driven trust dynamics in social networks. Procedia Soc. Behav. Sci. **79**, 27–37 (2013)

Heesacker, M., Petty, R.E., Cacioppo, J.T.: Field dependence and attitude change: source credibility can alter persuasion by affecting message-relevant thinking. J. Pers. **51**(4), 653–666 (1983)

Hegselmann, R., Krause, U.: Opinion dynamics and bounded confidence models, analysis, and simulation. J. Artif. Soc. Soc. Simul. **5**(3), (2002)

Herbert, T.T., Estes, R.W.: Improving executive decisions by formalizing dissent: the corporate devil's advocate. Acad. Manag. Rev. **2**(4), 662–667 (1977)

Holland, J.H.: Emergence: from chaos to order. OUP, Oxford (2000)

Hollingshead, A.B.: Communication, learning, and retrieval in transactive memory systems. J. Exp. Soc. Psychol. **34**(5), 423–442 (1998)

Horne, B.D., Adali, S..: This just in: fake news packs a lot in title, uses simpler, repetitive content in text body, more similar to satire than real news. In Eleventh International AAAI Conference on Web and Social Media (2017)

Hovland, C.I., Harvey, O.J., Sherif, M.: Assimilation and contrast effects in reactions to communication and attitude change. J. Abnorm. Soc. Psychol. **55**(2), 244–252 (1957)

Hutchins, E.: Distributed cognition. In: Neil, J.S., Paul, B.B. (eds.) The international encyclopedia of the social and behavioral sciences, pp. 2068–2072. Pergamon Press, Oxford (2001)

Hutchins, E.: The distributed cognition perspective on human interaction. In: Enfield, N.J., Levinson, S.C. (eds.) Roots of human sociality. culture, cognition and interaction, pp. 375–398. Oxford, Berg (2006)

Jacoby, J., Morrin, M., Jaccard, J., Gurhan, Z., Kuss, A., Maheswaran, D.: Mapping attitude formation as a function of information input: online processing models of attitude formation. J. Consum. Psychol. **12**(1), 21–34 (2002)

Janis, I.L.: Victims of groupthink: a psychological study of foreign-policy decisions and fiascoes. Houghton Mifflin, Boston (1972)

Janis, I.L.: Groupthink: psychological studies of policy decisions and fiascoes, vol. 349. Houghton Mifflin, Boston (1982)

Johnson, S.: Emergence: the connected lives of ants, brains, cities and software. Simon and Schuster, New York (2001)

Johnson, N.: Simply complexity: a clear guide to complexity theory. Oneworld Publications, Oxford (2009)

Johnson-George, C., Swap, W.C.: Measurement of specific interpersonal trust: construction and validation of a scale to assess trust in a specific other. J. Pers. Soc. Psychol. **43**(6), 1306 (1982)

Johnston, W.A., Dark, V.J., Jacoby, L.L.: Perceptual fluency and recognition judgments. J. Exp. Psychol. Learn. Mem. Cogn. **11**(1), 3 (1985)

Kacprzyk-Murawska, M.: Functional approach to trust, unpublished manuscpript. (2018)

Konovsky, M.A., Pugh, S.D.: Citizenship behavior and social exchange. Acad. Manag. J. **37**(3), 656–669 (1994)

Kramer, R.M., Tyler, T.R.: Trust in organizations: frontiers of theory and research. Sage, Thousand Oaks (1996)

Krosnick, J.A.: Attitude importance and attitude change. J. Exp. Soc. Psychol. **24**, 240–255 (1988)

Kruglanski, A.W.: Motivated social cognition: principles of the interface. (1996)

Kruglanski, A.W., Freund, T.: The freezing and unfreezing of lay-inferences: effects on impressional primacy, ethnic stereotyping, and numerical anchoring. J. Exp. Soc. Psychol. **19**(5), 448–468 (1983)

Kruglanski, A.W., Thompson, E.P.: Persuasion by a single route: a view from the unimodel. Psychol. Inq. **10**(2), 83–109 (1999)

Kruglanski, A.W., Webster, D.M.: Motivated closing of the mind: "Seizing" and "freezing". Psychol. Rev. **103**(2), 263 (1996)

Kruglanski, A.W., Pierro, A., Mannetti, L., De Grada, E.: Groups as epistemic providers: need for closure and the unfolding of group-centrism. Psychol. Rev. **113**(1), 84 (2006)

Kunda, Z., Thagard, P.: Forming impressions from stereotypes, traits, and behaviors: a parallel-constraint-satisfaction theory. Psychol. Rev. **103**(2), 284 (1996)

Landrum, A.R., Mills, C.M., Johnston, A.M.: When do children trust the expert? Benevolence information influences children's trust more than expertise. Dev. Sci. **16**(4), 622–638 (2013)

Latané, B., Nowak, A.: Attitudes as catastrophes: from dimensions to categories with increasing involvement. In: Dynamical systems in social psychology, pp. 219–249. Academic, San Diego (1994)

Lee, K.M., Nass, C.: The multiple source effect and synthesized speechdoubly-disembodied language as a conceptual framework. Hum. Commun. Res. **30**(2), 182–207 (2004)

Levine, J.M., Higgins, E.T.: Shared reality and social influence in groups and organizations. In: Social influence in social reality: promoting individual and social change, pp. 33–52. Hogrefe & Huber Publishers, Ashland (2001)

Lewicka, M., Czapiński, J., Peeters, G.: Positive-negative asymmetry or 'When the heart needs a reason'. Eur. J. Soc. Psychol. 22(5), 425–434 (1992)

Lewicki, R.J., Bunker, B.B.: Developing and maintaining trust in work relationships. In: Trust in organizations: frontiers of theory and research, pp. 114–139. Sage, Thousand Oaks (1996)

Lewicki, R., McAllister, D.J., Bies, R.: Trust and distrust: new relationships and realities. Acad. Manag. Rev. 23(3), 438–458 (1998)

Lewicki, R.J., Tomlinson, E.C., Gillespie, N.: Models of interpersonal trust development: theoretical approaches, empirical evidence, and future directions. J. Manag. 32(6), 991–1022 (2006)

Li, P.P.: Towards an interdisciplinary conceptualization of trust: a typological approach. Manag. Organ. Rev. 3(3), 421–445 (2007)

Liberman, N., Trope, Y.: The psychology of transcending the here and now. Science. 322(5905), 1201–1205 (2008)

Liberman, N., Trope, Y., Stephan, E.: Psychological distance. In: Social psychology: handbook of basic principles, 2, pp. 353-383 (2007)

Lorenz, J., Rauhut, H., Schweitzer, F., Helbing, D.: How social influence can undermine the wisdom of crowd effect. Proc. Natl. Acad. Sci. 108(22), 9020–9025 (2011)

Maddux, J.E., Rogers, R.W.: Effects of source expertness, physical attractiveness, and supporting arguments on persuasion: a case of brains over beauty. J. Pers. Soc. Psychol. 39(2), 235 (1980)

Mason, M., Hood, B., Macrae, C.N.: Look into my eyes: gaze direction and person memory. Memory. 12(5), 637–643 (2004)

Mayer, R.C., Davis, J.H., Schoorman, F.D.: An integrative model of organizational trust. Acad. Manag. Rev. 20(3), 709–734 (1995)

McCroskey, J., Burgoon, M.: Establishing predictors of latitude of acceptance-rejection and attitudinal intensity: a comparison of assumptions of social judgment and authoritarian personality theories. Speech Monogr. 41(4), 421–426 (1974). https://doi.org/10.1080/03637757409375868

McEvily, B., Tortoriello, M.: Measuring trust in organisational research: review and recommendations. J. Trust Res. 1(1), 23–63 (2011)

McGuire, W.J.: Attitudes and attitude change. In: The handbook of social psychology, pp. 233–346. Random House, New York (1985)

McKnight, D.H., Chervany, N.: While trust is cool and collected, distrust is fiery and frenzied: a model of distrust concepts. In AMCIS Proceedings, pp.883–888. (2001)

McKnight, D.H., Cummings, L.L., Chervany, N.L.: Initial trust formation in new organizational relationships. Acad. Manag. Rev. 23(3), 473–490 (1998)

McKnight, D.H., Kacmar, C.J., Choudhury, V.: Dispositional trust and distrust distinctions in predicting high and low-risk internet expert advice site perceptions. e-Service J. 3(2), 35–58 (2004)

McPherson, M., Smith-Lovin, L., Cook, J.M.: Birds of a feather: homophily in social networks. Annu. Rev. Sociol. 27, 415–444 (2001)

Milgram, S.: Obedience to authority. Harper & Row, New York (1974)

Misztal, B.: Trust in modern societies: the search for the bases of social order. Polity Press, Cambridge (1998)

Mitchell, R., Nicholas, S.: Knowledge creation in groups: the value of cognitive diversity, transactive memory and open-mindedness norms. Electron. J. Knowl. Manag. 4(1), 67–74 (2006)

Myers, D.G., Lamm, H.: The group polarization phenomenon. Psychol. Bull. 83(4), 602 (1976)

Noelle-Neumann, E.: The spiral of silence a theory of public opinion. J. Commun. 24(2), 43–51 (1974)

Nowak, A.: Dynamical minimalism: why less is more in psychology. Personal. Soc. Psychol. Rev. 8(2), 183–192 (2004)

Nowak, A., Vallacher, R.R.: Dynamical social psychology, vol. 647. Guilford Press, New York (1998)

Nowak, A., Szamrej, J., Latané, B.: From private attitude to public opinion: a dynamic theory of social impact. Psychol. Rev. **97**(3), 362 (1990)

Nowak, A., Vallacher, R.R., Read, S.J., Miller, L.C.: Toward computational social psychology: cellular automata and neural network models of interpersonal dynamics. In: Connectionist models of social reasoning and social behavior, pp. 277–311. Lawrence Erlbaum, Mahwah (1998)

Nowak, A., Vallacher, R.R., Tesser, A., Borkowski, W.: Society of self: the emergence of collective properties in self-structure. Psychol. Rev. **17**, 39–61 (2000)

Nowak, A., Vallacher, R.R., Miller, M.E.: Social influence and group dynamics. Handb. Psychol. 383–417 (2003)

Nowak, A., Winkowska-Nowak, K., Brée, D. (eds.): Complex human dynamics: from mind to societies. Springer, Berlin/Heidelberg (2013)

Nowak, A., Vallacher, R.R., Zochowski, M., Rychwalska, A.: Functional synchronization: the emergence of coordinated activity in human systems. Front. Psychol. **8**, 945 (2017)

Nowak, A., Ziembowicz K., Waszkiewicz J., Winkielman P.: Incoherence lowers trust. Unpublished manuscript (2018)

Nowak, A., Vallacher, R., Rychwalska, A., Zochowski, M.: In Sync: the emergence of function in minds, groups, and societies. Springer (2020)

Ormerod, P.: The economics of radical uncertainty (No. 2015-40). Economics Discussion Papers (2015)

Page, S.E.: The difference: how the power of diversity creates better groups, firms, schools, and societies. Princeton University Press, Princeton (2008)

Pariser, E.: The filter bubble: what the Internet is hiding from you. Penguin, London (2011)

Parks, C.D., Henager, R.F., Scamahorn, S.D.: Trust and reactions to messages of intent in social dilemmas. J. Confl. Resolut. **40**(1), 134–151 (1996)

Peeters, G., Czapiński, J.: Positive-negative asymmetry in evaluations: the distinction between affective and informational negativity effects. Eur. Rev. Soc. Psychol. **1**(1), 33–60 (1990)

Pennington, N., Hastie, R.: Explaining the evidence: tests of the story model for juror decision making. J. Pers. Soc. Psychol. **62**(2), 189 (1992)

Pennycook, G., Rand, D.G.: Lazy, not biased: susceptibility to partisan fake news is better explained by lack of reasoning than by motivated reasoning. Cognition. **188**, 39–50 (2019)

Petty, R.E., Cacioppo, J.T.: The elaboration likelihood model of persuasion. In: Communication and persuasion, pp. 1–24. Springer, New York (1986)

Putnam, R.D.: The prosperous community. Am. Prospect. **4**(13), 35–42 (1993)

Putnam, R.D.: Bowling alone: the collapse and revival of American community. Simon and Schuster, New York (2001)

PytlikZillig, L.M., Hamm, J.A., Shockley, E., Herian, M.N., Neal, T.M., Kimbrough, C.D., Tomkins, A.J., Bornstein, B.H.: The dimensionality of trust-relevant constructs in four institutional domains: results from confirmatory factor analyses. J. Trust Res. **6**(2), 111–150 (2016)

Reber, R., Winkielman, P., Schwarz, N.: Effects of perceptual fluency on affective judgments. Psychol. Sci. **9**(1), 45–48 (1998)

Reeder, G.D., Brewer, M.B.: A schematic model of dispositional attribution in interpersonal perception. Psychol. Rev. **86**(1), 61 (1979)

Riegelsberger, J., Sasse, M.A., McCarthy, J.D.: Do people trust their eyes more than their ears? Media bias while seeking expert advice. In Proceedings of CHI'05 extended abstracts on Human factors in Computing System. pp. 1745–1748 (2005)

Rokeach, M.: Political and religious dogmatism: an alternative to the authoritarian personality. Psychol. Monogr. Gen. Appl. **70**(18), 1 (1956)

Ross, L., Lepper, M.R., Hubbard, M.: Perseverance in self-perception and social perception: biased attributional processes in the debriefing paradigm. J. Pers. Soc. Psychol. **32**(5), 880 (1975)

Roszczyńska-Kurasińska, M., Kacprzyk-Murawska, M.: The dynamics of trust from the perspective of a trust game. In: Nowak, A., Winkowska-Nowak, K., Bree, D. (eds.) Complex human dynamics, pp. 191–207. Springer, Berlin/Heidelberg (2013)

Rotter, J.B.: Generalized expectancies for interpersonal trust. Am. Psychol. **26**(5), 443 (1971)

Rousseau, D.M., Sitkin, S.B., Burt, R.S., Camerer, C.: Not so different after all: a cross-discipline view of trust. Acad. Manag. Rev. **23**(3), 393–404 (1998)

Sawyer, R.K., Sawyer, R.K.S.: Social emergence: societies as complex systems. Cambridge University Press, Cambridge (2005)

Schachter, S.: Deviation, rejection, and communication. J. Abnorm. Soc. Psychol. **46**(2), 190 (1951)

Schoorman, F.D., Mayer, R.C., Davis, J.H.: An integrative model of organizational trust: past, present, and future. Acad. Manag. Rev. **32**(2), 344–354 (2007)

Schuman, H., Presser, S.: Questions and answers: experiments on question form, wording, and context in attitude surveys. Academic, New York (1981)

Schweiger, D.M., Sandberg, W.R., Rechner, P.L.: Experiential effects of dialectical inquiry, devil's advocacy and consensus approaches to strategic decision making. Acad. Manag. J. **32**(4), 745–772 (1989)

Schwenk, C.R.: Devil's advocacy in managerial decision-making. J. Manag. Stud. **21**(2), 153–168 (1984)

Schwenk, C.R.: Effects of devil's advocacy and dialectical inquiry on decision making: a meta-analysis. Organ. Behav. Hum. Decis. Process. **47**(1), 161–176 (1990)

Sherif, M.: The psychology of social norms. Harper, Oxford (1936)

Sherif, M., Sherif, C.: Acceptable and unacceptable behavior defined by group norms. Reference groups: exploration into conformity and deviation of adolescents. Harper & Row, New York (1964)

Sherif, C.W., Sherif, M., Nebergall, R.E.: Attitude and attitude change: the social judgement-involvement process. Saunders, Philadelphia/London (1965)

Simon, H.A.: Bounded rationality in social science: today and tomorrow. Mind & Society. **1**(1), 25–39 (2000)

Simon, D., Holyoak, K.J.: Structural dynamics of cognition: from consistency theories to constraint satisfaction. Personal. Soc. Psychol. Rev. **6**(4), 283–294 (2002)

Sitkin, S.B., Roth, N.L.: Explaining the limited effectiveness of legalistic "remedies" for trust/distrust. Organ. Sci. **4**(3), 367–392 (1993)

Snow, C.P.: The two cultures and the scientific revolution. Cambridge University Press, New York (1959)

Stephan, E., Liberman, N., Trope, Y.: Politeness and psychological distance: a construal level perspective. J. Pers. Soc. Psychol. **98**(2), 268–280 (2010)

Strogatz, S.H.: Exploring complex networks. Nature. **410**(6825), 268 (2001)

Sunstein, C.R., Hastie, R.: Making dumb groups smarter. Harv. Bus. Rev. **92**(12), 90–98 (2014)

Tang, L.R., Jang, S.S., Chiang, L.L.: Website processing fluency: its impacts on information trust, satisfaction, and destination attitude. Tour. Anal. **19**(1), 111–116 (2014)

Thagard, P.: Explanatory coherence. Behav. Brain Sci. **12**(03), 435–467 (1989)

Thagard, P.: Coherence in thought and action. MIT press, Cambridge (2002)

Thibaut, J.W., Kelley, H.H.: The social psychology of groups. Routledge, London (2017/1959)

Thom, R.: Structural stability and morphogenesis. Benjamin, New York (1975)

Thurstone, L.L.: Multiple factor analysis. University of Chicago Press, Chicago (1947)

Tyler, T.R.: Trust within organisations. Pers. Rev. **32**(5), 556–568 (2003)

Ulrich, H., Probst, G.J. (eds.): Self-organization and management of social systems: insights, promises, doubts, and questions (Vol. 26). Springer (2012)

Uviller, H.R.: Credence, character, and the rules of evidence: seeing through the liar's tale. Duke Law J. **42**, 776–832 (1993)

Uzzi, B., Dunlap, S.: How to build your network. Harv. Bus. Rev. **83**(12), 53 (2005)

Vallacher, R.R., Nowak, A. (eds.): Dynamical systems in social psychology. Academic, San Diego (1994)

Vallacher, R.R., Nowak, A.: The dynamics of self-regulation. In: Advances in social cognition, vol. 12, pp. 3–52. Lawrence Erlbaum Associates, Mahwah (1999)

Vallacher, R.R., Nowak, A.: Dynamical social psychology: finding order in the flow of human experience. In: Social psychology: handbook of basic principles, vol. 2, pp. 734–758. Guilford Press, New York (2007)

Vallacher, R.R., Wegner, D.M.: A theory of action identification. Erlbaum, Hillsdale (1985)

Vallacher, R.R., Wegner, D.M.: What do people think they're doing? Action identification and human behavior. Psychol. Rev. **94**(1), 3 (1987)

Vallacher, R.R., Read, S.J., Nowak, A.: The dynamical perspective in personality and social psychology. Personal. Soc. Psychol. Rev. **6**(4), 264–273 (2002)

Vallacher, R.R., Read, S.J., Nowak, A. (eds.): Computational social psychology. Routledge, New York (2017)

Voci, A.: The link between identification and in-group favouritism: effects of threat to social identity and trust-related emotions. Br. J. Soc. Psychol. **45**(2), 265–284 (2010). https://doi.org/10.1348/014466605X52245

Vosoughi, S., Roy, D., Aral, S.: The spread of true and false news online. Science. **359**(6380), 1146–1151 (2018)

Watts, D.J., Strogatz, S.H.: Collective dynamics of 'small-world'networks. Nature. **393**(6684), 440 (1998)

Wegner, D.M.: Transactive memory: a contemporary analysis of the group mind. In: Mullen, B., Goethals, G.R. (eds.) Theories of group behavior, pp. 185–208, New York. Springer (1987)

Wegner, D.M., Vallacher, R.R.: Implicit psychology: an introduction to social cognition. Oxford University Press, Oxford (1977)

Weisbuch, G., Deffuant, G., Amblard, F., Nadal, J.P.: Interacting agents and continuous opinions dynamics. In: Heterogenous agents, interactions and economic performance, pp. 225–242. Springer, Berlin Heidelberg (2003)

Wickham H. (2009) ggplot2: Elegant Graphics for Data Analysis. Springer-Verlag, New York

Winkielman, P., Olszanowski, M., Gola, M.: Faces in-between: evaluations reflect the interplay of facial features and task-dependent fluency. Emotion. **15**(2), 232 (2015)

Wojciszke, B.: Postawy i ich zmiana. Psychologia. **3**, 79–105 (2000)

Wojciszke, B.: Morality and competence in person-and self-perception. Eur. Rev. Soc. Psychol. **16**(1), 155–188 (2005)

Wojciszke, B., Baryła, W.: Perspektywa sprawcy i biorcy w spostrzeganiu siebie i innych. Psychol. Społeczna. **1**(1), 9–32 (2006)

Wojciszke, B., Bazińska, R., Jaworski, M.: On the dominance of moral categories in impression formation. Personal. Soc. Psychol. Bull. **24**(12), 1251–1263 (1998)

Worchel, P.: Trust and distrust. In: Austin, W.G., Worchel, S. (eds.) The social psychology of intergroup relations. Wadsworth, Belmont (1979)

Yoo, Y., Kanawattanachai, P.: Developments of transactive memory systems and collective mind in virtual teams. Int. J. Organ. Anal. **9**(2), 187–208 (2001)

Zabłocka, A., Praszkier, R., Petrushak, E., Kacprzyk-Murawska, M.: Measuring the propensity for building social capital depending on ties-strength. J. Posit. Manag. **7**(4), 19–39 (2017)

Ziegler, C.N.: On recommender systems. In social web artifacts for boosting recommenders, pp. 11–20. Springer, Cham (2013)

Zucker, L.G., Darby, M.R., Brewer, M.B., Peng, Y.: Collaboration structure and information dilemmas in biotechnology: organizational boundaries and trust production. In: Kramer, R.M., Tyler, T.R. (eds.) Trust in organizations: frontiers of theory and research, pp. 90–113. Sage, Thousand Oaks (1996)

Index

Group roles, 44, 65–66
Groupthink, 23, 44, 53–55, 57, 63, 66, 67

H
High-level information, 20, 21, 23

I
Importance, 2, 3, 7–12, 18, 20–24, 28, 34, 43,
    51, 53, 55, 57, 63, 66, 70
Incoherence, 16–21, 24, 26, 27, 29–34, 39, 40,
    49, 52–54, 63, 65, 66, 70
Influence/advice accepting, 7, 16, 19, 21, 50
Information processing systems, 2, 3, 22,
    43–68
Intentions, 4, 7, 11, 13, 14, 23, 25, 26
Interpersonal coherence, 30, 31
Intrapersonal coherence, 30, 39

J
Judgment/judgments, 1–3, 5, 6, 8–11, 13–21,
    23–26, 32, 39, 46, 48, 54, 55, 66–70

L
Latitude of acceptance/rejection, 48, 49,
    51–53
Levels of incoherence, 16, 19, 21
Low-level information, 23

M
Morality, 9, 13, 25, 26

N
Need for cognitive closure, 10

O
Opinion dynamics, 10, 43, 50–54, 64
Opinions/opinion, 1–4, 6, 9, 10, 17, 19, 20,
    22–24, 31, 34–40, 43–68
Optimality, 3, 22, 44, 45, 56, 57
Own expertise, 17, 19, 21

P
Peripheral route processes, 10
Personal relevance, 9
Physical distance, 9
Positive/negative asymmetry, 15
Predictability, 25, 60
Propensity of trust, 26, 28, 45, 47, 49

R
Readiness to make decision, 2, 29–31, 39
Responsibility, 5, 7, 25, 26

S
Shared reality, 2, 3, 44, 46, 47, 53
Small world network, 49
Social capital, 6, 12, 44, 65, 70
Social groups, 2, 3, 5, 6, 23, 64
Social identity, 9, 13
Social influence, 1–24, 43–45, 48, 49, 53, 63,
    65, 66, 69, 70
Social judgment, 13–15, 17, 25, 48
Social media, 67
Social proof, 47
Source of influence/influencing agent, 2–4,
    6–8, 10, 11, 13, 17, 52
Spiral of silence, 68

T
Target of influence, 2–7
Temporal distance, 9
Transactive memory, 6, 8, 22, 46
Trust, 1, 3, 8, 11–32, 34–40, 44–64, 66, 67, 70
Trustfulness, 44, 45, 47, 49–52, 54, 63
Trust game, 26, 27, 29
Trustworthiness, 11, 15, 26, 28, 30, 39, 45–47,
    49, 52, 53, 55

U
Unimodel, 8, 10

W
Willing/willingness/to help, 1, 11, 13, 21, 26,
    29, 39, 43, 49
Wisdom of the crowds, 66, 67